Springer Tracts in Mechanical Engineering

Springer Tracts in Mechanical Engineering (STME) publishes the latest developments in Mechanical Engineering - quickly, informally and with high quality. The intent is to cover all the main branches of mechanical engineering, both theoretical and applied, including:

- Engineering Design
- Machinery and Machine Elements
- Mechanical structures and stress analysis
- Automotive Engineering
- Engine Technology
- Aerospace Technology and Astronautics
- Nanotechnology and Microengineering
- Control, Robotics, Mechatronics
- MEMS
- Theoretical and Applied Mechanics
- Dynamical Systems, Control
- Fluids mechanics
- Engineering Thermodynamics, Heat and Mass Transfer
- Manufacturing
- Precision engineering, Instrumentation, Measurement
- Materials Engineering
- Tribology and surface technology

Within the scopes of the series are monographs, professional books or graduate textbooks, edited volumes as well as outstanding Ph.D. theses and books purposely devoted to support education in mechanical engineering at graduate and post-graduate levels.

Indexed by SCOPUS and Springerlink.

To submit a proposal or request further information, please contact: Dr. Leontina Di Cecco Leontina.dicecco@springer.com or Li Shen Li.shen@springer.com.

Please check our Lecture Notes in Mechanical Engineering at http://www.springer.com/series/11236 if you are interested in conference proceedings. To submit a proposal, please contact Leontina.dicecco@springer.com and Li.shen@springer.com.

More information about this series at http://www.springer.com/series/11693

Esteban Ferrer · Adeline Montlaur
Editors

Recent Advances in CFD for Wind and Tidal Offshore Turbines

Springer

Editors
Esteban Ferrer
ETSIAE-UPM - School of Aeronautics
Universidad Politécnica de Madrid
Madrid, Spain

Adeline Montlaur
Escola d'Enginyeria de Telecomunicació i
Aeroespacial de Castelldefels
Universitat Politècnica de Catalunya
Castelldefels, Barcelona, Spain

ISSN 2195-9862 ISSN 2195-9870 (electronic)
Springer Tracts in Mechanical Engineering
ISBN 978-3-030-11886-0 ISBN 978-3-030-11887-7 (eBook)
https://doi.org/10.1007/978-3-030-11887-7

Library of Congress Control Number: 2018968533

This Springer imprint is published by the registered company Springer Nature Switzerland AG
The registered company address is: Gewerbestrasse 11, 6330 Cham, Switzerland

Contents

Introduction

Offshore wind and tidal turbines are evolving as important renewable technologies to supply clean energy. In particular, the unpredictability of wind or solar energy can be compensated by the predictability of tidal turbines, which are governed by periodic and predictable tidal cycles. Consequently, tidal energy provides a constant reliable source of clean energy. Regarding offshore wind, the less turbulent atmospheric boundary layer on the sea surface (when compared to onshore sites) has the potential for increasing energy harvesting. Furthermore, less restrictive regulations in offshore environments have enabled larger structures and turbine rotors. Offshore wind turbines can be fixed to the sea ground in shallow waters or can float in deeper waters.

Tidal turbines share aerodynamic characteristics with their close relative wind turbines. Both technologies use airfoil-shaped blades that rotate driven by the lift force produced by the flow on the blades. The environments in which wind and tidal turbines operate share some similarities, and both environments experience high levels of turbulence with eddies of variable size and shape, and thick boundary layers shaping the incoming flow. However, some environmental conditions are characteristic of the tidal environments. The presence of the deformable sea surface, its confining effect and the deformation due to the energy extracted by tidal turbines are major differences between air and water technologies. Among the types of offshore turbines that are being considered by manufactures, one can distinguish between *horizontal axis turbines* (HAT) or *axial flow turbines* (AFT) and *vertical axis turbines* (VAT) or *crossflow turbines* (CFT).

On the one hand, horizontal axis turbines (HATs or AFTs) have their axis of rotation aligned with the flow stream. This technology requires orientation for the rotor plane to be perpendicular to the flow stream. Tidal devices have simpler orientation mechanisms since tidal currents are mainly bidirectional (tidal cycle) and 180° blade pitching suffices. In marine devices, blades are thicker and rotate more slowly to withstand the augmented forces resulting from the higher density of the water. On the other hand, vertical axis turbines (VATs or CFTs) have their axis of rotation perpendicular to the flow stream. This type of device has the advantage of not requiring orientation since they rotate independently of the stream direction and hence some researchers argue that this technology is more suitable for offshore environment as it would minimise maintenance costs through reduced control systems (e.g. yaw mechanism). Their main drawback is that they are generally less efficient than HAT. To date, VATs have had limited use within the wind energy sector, where the three-bladed axial flow turbine (HAT) has been widely adopted. However, it is thought that VAT configurations may be advantageous for new emergent markets as offshore wind and tidal turbines.

In summary, the simulation of offshore turbines requires taking into account some, if not all of the complex physics of offshore environments (e.g. sea surface deformation) and different turbine types, and consequently, new CFD tools are required. In 2015, we published "CFD for Wind and Tidal Offshore Turbines" [1], where various authors contributed to new CFD techniques to tackle some of the physics of offshore turbines. The present book incorporates more recent developments in the topic.

This book encompasses novel CFD techniques to compute offshore wind and tidal applications. All included papers have been presented at the *6th European Conference on Computational Mechanics (Solids, Structures and Coupled Problems) (ECCM 6) and the 7th European Conference on Computational Fluid Dynamics (ECFD 7)* that was held in Glasgow in 2018. The book includes contributions of researchers from academia and industry.

Madrid, Spain
December 2018

Esteban Ferrer
Adeline de Montlaur

Reference

1. Ferrer E, Montlaur A (2015) CFD for wind and tidal offshore turbines. Springer tracts in mechanical engineering, Springer Int Publishing AG

Simple Models for Cross Flow Turbines

Esteban Ferrer and Soledad Le Clainche

Abstract Using a high order discontinuous Galerkin numerical method with sliding meshes, we simulate one, two and three bladed cross-flow turbines to extract statistics of the generated wakes (time averaged velocities and Reynolds stresses). Subsequently, we compare the wakes resulting from simple models (a circular cylinder and an actuator disc) to the time averaged cross-flow turbine wakes. Additionally, we provide results for a reduced order model based on dynamic mode decomposition (Le Clainche and Ferrer, Energies, 11(3), 2018, [1]). Whilst simplified models find difficulties in capturing wake asymmetries characteristic of cross-flow turbines, our proposed reduced order model captures mean values and Reynolds stresses with good accuracy, showing the potential of the last technique to speed up the simulation of cross-flow turbine statistics.

1 Introduction

Cross Flow Turbines (CFT) are also referred to as vertical-axis, H-rotors or Darrieus type turbines, however the term cross-flow turbine is preferred since the absolute turbine position is omitted and the relative flow-axis geometry is emphasised through this terminology. This type of turbine has had limited use within the onshore wind energy sector, where the three bladed axial flow concept has been widely adopted. However, it has been argued (e.g. [2, 3]) that CFT may be advantageous for wind and tidal offshore environments. On the one hand, these turbines do not require orientation as they rotate independently of the stream direction, which leads to reduced control systems (e.g. yaw mechanism) and diminished maintenance costs. On the second hand, these devices allow for the generator to be located at sea level (and not in the

E. Ferrer (✉) · S. Le Clainche
ETSIAE-UPM - School of Aeronautics, Universidad Politécnica de Madrid,
Plaza Cardenal Cisneros 3, 28040 Madrid, Spain
e-mail: esteban.ferrer@upm.es

S. Le Clainche
e-mail: soledad.leclainche@upm.es

E. Ferrer and A. Montlaur (eds.), *Recent Advances in CFD for Wind and Tidal Offshore Turbines*, Springer Tracts in Mechanical Engineering,
https://doi.org/10.1007/978-3-030-11887-7_1

nacelle at height, as in horizontal axis turbines), which enables the design of large turbines, suitable for offshore sites.

Cross-flow turbines consist of foil shaped blades that generate lift forces so as to rotate a shaft to which the blades are connected. Therefore azimuthal changes in blade aerodynamics (or hydrodynamics)[1] are common, resulting in complex flow phenomena such as stalled flows, vortex shedding and blade-vortex interactions, e.g. [4–6]. These complex features modify the turbine behaviour and shape its wake, see for example [7].

The simulation and prediction of the space-time evolution of flows around CFT is difficult. This difficulty is caused, on the one hand, by the above-mentioned blade motion, which leads to unsteady flow features and dynamic effects, but also to the turbulent flow regime in which these devices operate (i.e. high Reynolds numbers). Both the relative foil movement and the turbulent regime lead typically to costly numerical simulations, such as the large eddy simulations with sliding meshes included in this text.

Various approaches exist to avoid the expensive computational cost of resolving blade motion and turbulent flows. Popular examples include modelling the turbine (and blade action) using a porous disc, also called actuator discs [8, 9], instead of representing the geometry in detail, or replacing moving blades by static objects that provide similar averaged wakes, see [10] where a cylinder was proposed to replace the CFT. In this work, we explore these simple models and compare their time averaged wakes to the ones issued from the rotating high order Large Eddy Simulations (LES). In addition, we propose the use of a reduced order model [1] based on high order Dynamic Mode Decomposition to represent the turbine wakes.

We begin by simulating one, two and three bladed turbines using a high order discontinuous Galerkin solver combined with a LES approach [1, 11–14]. These simulations are used to characterise the one, two and three bladed turbine wakes in terms of time averaged velocities and Reynolds stresses to subsequently compare simple models that avoid the necessity for costly computations.

2 Cross-Flow Turbine Simulations

We start by simulating one, two and three bladed turbines using a high order (≥ 2) numerical method developed by the first author [1, 11–15]. This high order solver provides highly accurate solutions on static and moving meshes composed of mixed triangular-quadrilateral meshes and can cope with curved boundary elements.

High order discontinuous Galerkin (DG) methods are characterised by low numerical errors (i.e., dispersion and diffusion) and their ability to use mesh refinement

[1] Airfoil aerodynamics and hydrofoil hydrodynamics are equivalent nomenclatures for foils operating in air or water environments. Since this work encompasses both wind and tidal turbine applications, from this point onwards, "foils" will denote either "airfoils" or "hydrofoils". In addition, the term "aerodynamic" can always be replaced by "hydrodynamic" in this work.

(H-refinement) and/or polynomial enrichment (P-refinement) in order to achieve accurate solutions [16]. Polynomial enrichment provides exponential decay of the error for smooth solutions as opposed to the typical constant decay provided by the H-refinement strategy. The P-refinement strategy enables accurate solutions with fewer number of degrees of freedom.

In this work, flow solutions of the nonlinear incompressible Navier–Stokes (NS) equations, are obtained from the 3D unsteady high order (order ≥ 2) H/P Discontinuous Galerkin—Fourier solver developed by the first author [1, 11–15]. The solver uses a second order stiffly stable approach to discretise the NS equations in time whilst spatial discretisation is provided by the discontinuous Galerkin—Symmetric Interior Penalty formulation with modal basis functions in the x-y plane. Here, x represents the streamwise flow direction and y is the normal direction (with respect to the axis of rotation). Spatial discretisation in the z-direction (here defining the spanwise turbine length) is provided by a purely spectral method that uses Fourier series and allows computation of spanwise periodic three-dimensional flows. Since high order methods (e.g. discontinuous Galerkin and Fourier) are unable to provide enough numerical dissipation to enable under-resolved high Reynolds computations (e.g. as necessary in LES), we have adapted the original laminar version of the solver to increase (controllably) the dissipation and enhance the stability in under-resolved simulations [15]. This dissipative formulation has minimal impact on well resolved flow regions and its implicit treatment does not restrict the use of relatively large time steps, thus providing an efficient stabilization mechanism for LES. The resulting solver enables 3D LES of rotating vertical axis turbines [1, 15].

2.1 High Order Solver Validations

The solver has been widely validated for a variety of flows, including bluff body flows, airfoil and blade aerodynamics under static and rotating conditions [11–13], global instability analysis [17, 18] and turbulent regimes [1, 15]. The advantages of high order methods (over low order) to compute turbine flows have been discussed in [19].

Validation for one bladed rotating turbine under turbulent regime were presented in [1, 15], where it was shown that blade forces are well captured by the high order solver. In the latter work, comparisons with a commercial low order solver (Ansys-Fluent) showed that the high order solver provides accurate forces and better resolution of the underlying flow physics in CFTs. Here, in Fig. 1, we include a comparisons with experiments [20] for a rotating one bladed turbine (extracted from [1, 15]). Flow conditions are detailed in Table 1. It can be seen that the simulated normal force agrees remarkably well with the experimental data, which includes pressure and strain gauge measurements (the latter being more accurate, as reported in [20]).

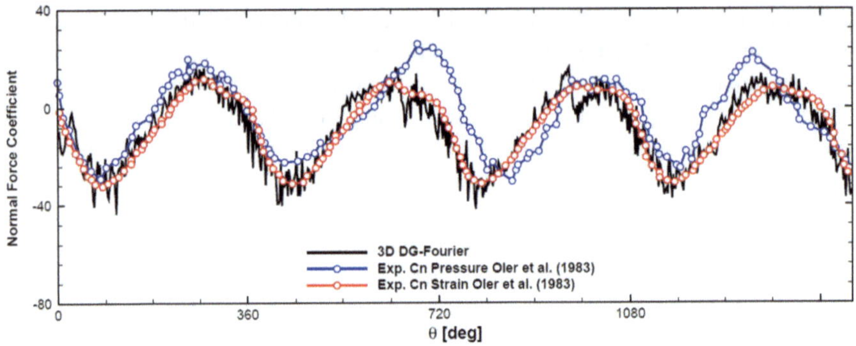

Fig. 1 Normal force evolution for a one bladed CFT turbine. The figure compares high order simulations (see details in [1, 15]) to experimental data [20]

Table 1 Vertical axis turbine conditions. All in international metric system. $TSR = \omega D/2U$ denotes the tip-speed ratio, $Re_c = \frac{c}{\nu}\sqrt{(\omega D/2)^2 + U^2}$ is the Reynolds number based on the chord and rotating speed whilst $Re_D = \frac{DU}{\nu}$ is based on the turbine diameter and free stream velocity

	Blade chord	Turb. diameter	Length ratio	Stream vel.	Rot. speed	Tip speed ratio	Kin. Visc	Reynolds	
Symbol	c	D	c/D	U	ω	TSR	ν	Re_c	Re_D
Units	m	m	–	m/s	rad/s	–	m^2/s	–	–
Simulations	1.0	8.0	0.125	0.088	0.11	5	5.0×10^{-5}	0.90×10^4	0.14×10^5

2.2 One, Two and Three Bladed Turbines

We simulate 3D LES flows for one, two and three bladed turbines at the same flow conditions and compare the issued wakes. The flow conditions are summarised in Table 1. All turbines use NACA0015 airfoils for the definition of their blades. Meshes used for the simulations use 0.4–1 million degrees of freedom. The mesh and the plane where the statistics are collected at $x/D = 1$ are shown in Fig. 2. Statistics are collected in all cases for not less than 5 turbine revolutions.

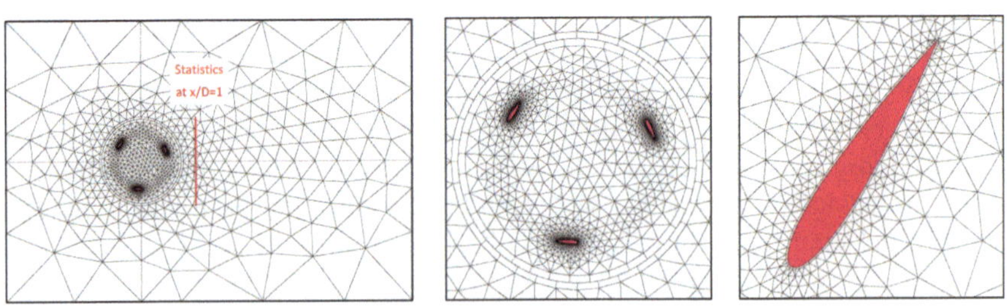

Fig. 2 Mesh used for the 3 bladed CFT simulation (using a polynomial order 3), the figure includes the plane at $x/D = 1$ where the statistics are collected (in red)

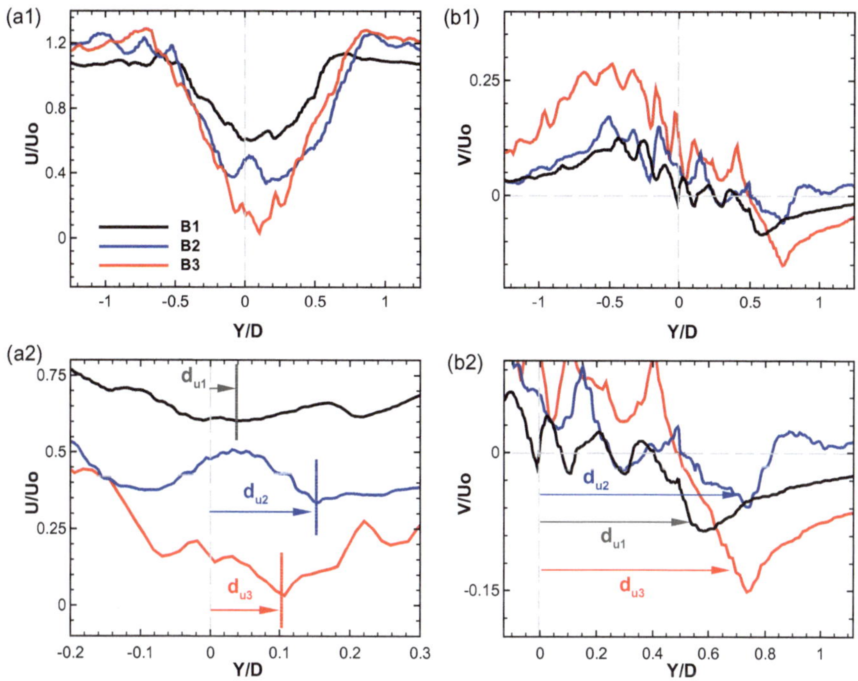

Fig. 3 Mean streamwise (U) and transversal (V) velocity components at $x/D = 1$ for one, two and three bladed turbines (B1, B2 and B3, respectively). Bottom figure includes the deviances from the horizontal axis d_{u1}, d_{u2}, d_{u3} and d_{v1}, d_{v2}, d_{v3}, where the sub-index u, v indicates the velocity component and the number defines one, two or three bladed turbine

We first compare the mean streamwise U and transverse V flow velocity components (top left and right, respectively) for the one, two and three turbines (B1, B2 and B3 respectively) in Fig. 3. We observe a more important velocity deficit when the number of blades is increased. The asymmetry of the transverse velocity V is also increased for larger number of blades, which is visible, particularly, in the blade retracting region ($y/D < 0$). To characterise the asymmetry of the wake, we calculate the deviance from the horizontal axis in the bottom figures and denote the deviances: d_{u1}, d_{u2}, d_{u3} and d_{v1}, d_{v2}, d_{v3}, where the sub-index u, v indicates the velocity component and the number defines one, two or three bladed turbine. We can quantify the deviance from symmetry and obtain: $d_{u1}/D = 0.03$, $d_{u2}/D = 0.15$, $d_{u3}/D = 0.1$ and $d_{v1}/D = 0.5$, $d_{v2}/D = 0.7$, $d_{v3}/D = 0.7$. It is interesting to note that the deviances are not located at the same asymmetric distance for the streamwise and the transverse velocity components, but that in all cases, the 3 bladed turbine almost doubles the deviance of the one bladed turbine. We conclude that the wakes arising from CFT turbines are far from symmetric and that asymmetries increase when increasing the number of blades. Similar asymmetries for three bladed CFTs have been reported in the recent work of Ouro et al. [21].

3 Simple Models for Cross-Flow Turbines

In this section, we compare the statistics obtained with the high order solver and sliding meshes (moving blades) to simple models that do not include moving parts and are cheaper to run.

3.1 Comparisons to a Circular Cylinder

In this section, we compare the wake statistics of rotating turbines to a non-rotating circular cylinder at $Re = 3900$. Results for the cylinder have been validated to experiential data and DNS in [15], showing very good agreement.

Figure 4 shows the mean streamwise and transverse velocities for the rotating turbines and the cylinder. The good agreement between the three bladed turbine (B3) and the cylinder is remarkable for both mean velocity components. However, since the cylinder wake is symmetric with respect to the horizontal position y, there is a consistent discrepancy in the comparisons. As noted in the previous section, wake asymmetry is embedded in CFT wakes and is not captured with this simple model. In addition, we compare the Reynolds stresses (i.e. time averaged $u'u'$ and $u'v'$) obtained between the three bladed turbine and the cylinder, in Fig. 5. In this case, the asymmetries are more pronounced, showing that the cylinder is not an appropriate model to mimic CFT at the proposed flow conditions.

3.2 Comparisons to a Porous Disc

In this section, we compare mean velocity components to porous disc simulations. The porous disc pressure jump is adjusted through the expression $\Delta p = -C\frac{1}{2}\rho v^2$,

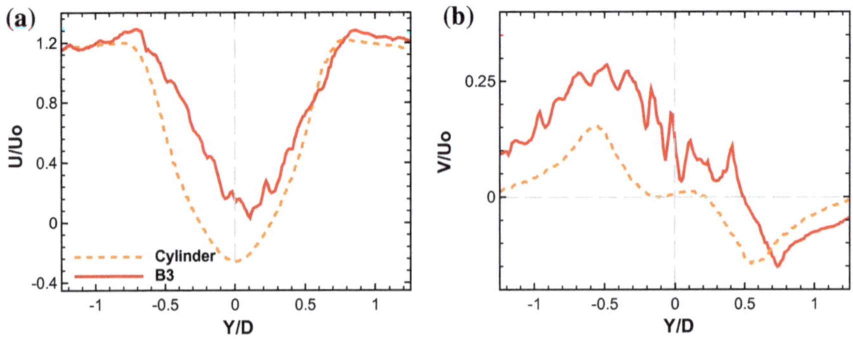

Fig. 4 Mean streamwise (U) and transversal (V) mean velocity components at $x/D = 1$ for circular cylinder compared to the wake of the three bladed turbine B3

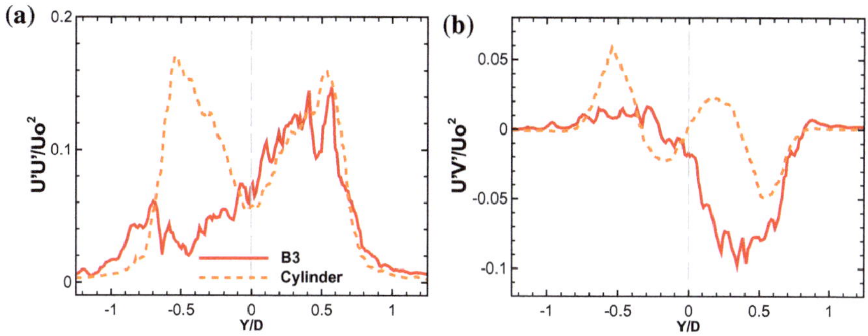

Fig. 5 Reynolds stresses at $x/D = 1$ for circular cylinder compared to the wake of the three bladed turbine B3

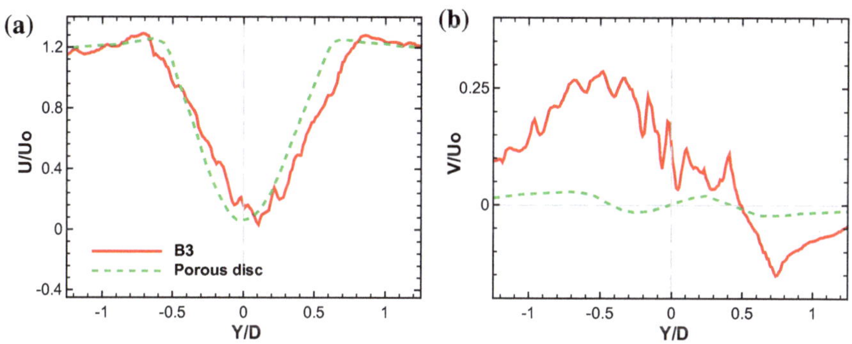

Fig. 6 Mean streamwise (U) and transversal (V) mean velocity components at $x/D = 1$ for a porous disc compared to the wake of the three bladed turbine B3

where $C(y)$ is a function that precludes the pressure with respect to the horizontal position y, and ρ, v are fluid density and velocity at the disc. Adjusting $C(y)$ we obtain a better match to the mean streamwise velocity component (i.e. triangular wake shape), that when using a unique constant value for C (i.e. rectangular wake shape), independent of the vertical position y.

Results for the mean velocities are shown in Fig. 6. It can be seen that whilst the streamwise velocity is well captured through this technique, the transverse component is very different, missing the physics associated to CFTs. Not being able to capture the mean components, there is no chance that the Reynolds stresses issued from the porous disc match the high order simulations (these last comparisons are not shown).

3.3 Reduced Order Model Based on High Order Dynamic Mode Decomposition

In [1], the authors proposed a Reduced Order Model (ROM) based on snapshots extracted from accurate high order simulations, to the authors' knowledge, this was the first ROM proposed for CFTs. Our ROM, based on Higher Order Dynamic Mode Decomposition (HODMD) [22–24], extracts information of the flow dynamics of a dynamical system (here the cross flow turbine), to enable the characterisation of the system through selected modes associated to the system frequencies. By selecting the most relevant frequencies, see [1], it is possible to generate a ROM that can mimic the flow evolution in space and time, but that also reproduces wake statistics.

To obtain the ROM, we simulate the rotating turbine for less than one revolution (an azimuthal angle of 270°) and extract snapshots of the flow field as time evolves. These snapshots, that include all velocity components, are used to generate the ROM. Once the ROM is generated by selecting appropriate modes issued from the high order Dynamic Mode Decomposition method, we generate statistics behind the turbine and compare the ROM statistics to the LES statistics obtained for the three bladed CFT. The advantage of the ROM method is twofold. First, to create the ROM we only need to simulate less than one rotation of the turbine (to generate snapshots). Second, once the ROM has been created, we can generate statistics for 5 rotations in less than 30 seconds using the ROM. In this case, the ROM was generated using 153 snapshots and only 5 modes where retained. Figure 7 shows the comparison for the mean streamwise velocity component U and Reynolds stresses $u'v'$. We observe reasonable good agreement between the ROM and the simulation. In particular, the ROM captures wake asymmetries both in the mean velocity and Reynolds stresses.

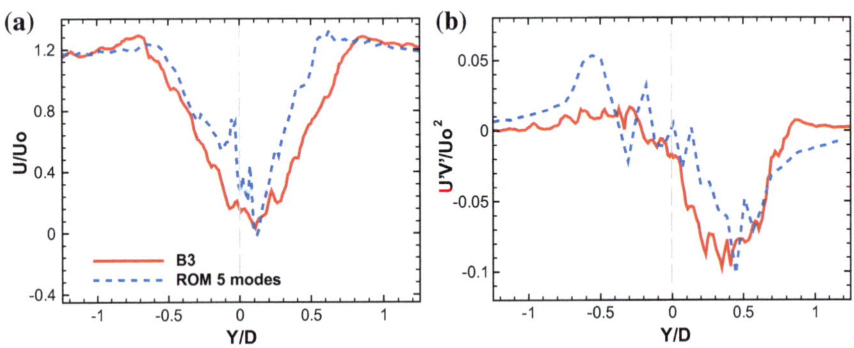

Fig. 7 **a** Mean streamwise velocity component (U) and **b** Reynolds stresses $u'v'$ at $x/D = 1$ for the proposed ROM compared to the wake of the three bladed turbine B3

4 Conclusions

We have simulated one, two and three bladed cross flow turbines using a high order LES solver with sliding meshes. This solver enables to capture time and space variations of the blades and is used to calculated flow statistics behind CFTs.

First, we show that all CFT wakes (one, two or three blades) are asymmetric with respect to their horizontal axis. Second, we compare wakes resulting from simple model (circular cylinder and porous discs) to the simulated wakes. We show that these simple models may give similar overall mean velocity distributions but are not able to capture the asymmetries characteristics of CFT. Finally, we generate wake statistics using the ROM methods described in [1]. The ROM model (which is cheap to run) is able to capture wake asymmetries and agrees well with the three bladed CFT wake. This promising technique will be further explored in future work.

References

1. Le Clainche S, Ferrer E (2018) A reduced order model to predict transient flows around straight bladed vertical axis wind turbines. Energies 11(3)
2. Gretton GI, Bruce T (2005) Preliminary results from analytical and numerical models of a variable-pitch vertical-axis tidal current turbine. In: 6th European wave and tidal energy conference, Glasgow, UK, September 2005
3. Gretton GI, Bruce T (2006) Hydrodynamic modelling of a vertical-axis tidal current turbine using a Navier–Stokes solver. In: Proceedings of the 9th world renewable energy congress, Florence, Italy, 2006
4. Ferrer E, Willden RHJ (2015) Blade-wake interactions in cross-flow turbines. Int J Mar Energy 11:71–83
5. Montlaur A, Giorgiani G (2015) Numerical study of 2D vertical axis wind and tidal turbines with a degree-adaptive hybridizable discontinuous Galerkin Method. In: Ferrer E, Montlaur A (eds) CFD for wind and tidal offshore turbines, Chap 2. Springer Tracts in Mechanical Engineering. Springer, Cham pp 13–26
6. Somoano M, Huera-Huarte FJ (2017) Flow dynamics inside the rotor of a three straight bladed cross-flow turbine. Appl Ocean Res 69:138–147
7. Bachant P, Wosnik M (2015) Characterising the near-wake of a cross-flow turbine. J Turbul 16(4):392–410
8. Islam M, Ting DSK, Fartaj A (2008) Aerodynamic models for Darrieus-type straight-bladed vertical axis wind turbines. Renew Sustain Energy Rev 12(4):1087–1109
9. Newman BG (1983) Actuator-disc theory for vertical-axis wind turbines. J Wind Eng Ind Aerodyn 15(1–3):347–355
10. Araya DB, Colonius T, Dabiri JO (2017) Transition to bluff-body dynamics in the wake of vertical-axis wind turbines. J Fluid Mech 813:346–381
11. Ferrer E (2012) A high order Discontinuous Galerkin - Fourier incompressible 3D Navier–Stokes solver with rotating sliding meshes for simulating cross-flow turbines. PhD thesis, University of Oxford, 2012
12. Ferrer E, Willden RHJ (2011) A high order discontinuous Galerkin finite element solver for the incompressible Navier–Stokes equations. Comput Fluids 46(1):224–230
13. Ferrer E, Willden RHJ (2012) A high order discontinuous Galerkin - Fourier incompressible 3D Navier–Stokes solver with rotating sliding meshes. J Comput Phys 231(21):7037–7056

14. Ferrer E, Moxey D, Willden RHJ, Sherwin S (2014) Stability of projection methods for incompressible flows using high order pressure-velocity pairs of same degree: continuous and discontinuous Galerkin formulations. Commun Comput Phys 16(3):817–840
15. Ferrer E (2017) An interior penalty stabilised incompressible discontinuous Galerkin - Fourier solver for implicit large eddy simulations. J Comput Phys 348:754–775
16. Wang ZJ, Fidkowski K, Abgrall R, Bassi F, Caraeni D, Cary A, Deconinck H, Hartmann R, Hillewaert K, Huynh HT, Kroll N, May G, Persson PO, van Leer B, Visbal M (2013) High-order CFD methods: current status and perspective. Int J Numer Methods Fluids 72(8):811–845
17. Ferrer E, de Vicente J, Valero E (2014) Low cost 3D global instability analysis and flow sensitivity based on dynamic mode decomposition and high-order numerical tools. Int J Numer Methods Fluids 76(3):169–184
18. Gonzalez LM, Ferrer E, Diaz-Ojeda HR (2017) Onset of three-dimensional flow instabilities in lid-driven circular cavities. Phys Fluids 29(6):064102
19. Ferrer E, Le Clainche S (2015) Flow scales in cross-flow turbines. In: Ferrer E, Montlaur A (eds) CFD for wind and tidal offshore turbines, Chap1. Springer tracts in mechanical engineering. Springer, Cham, pp 1–11
20. Oler JW, Strickland JH, Im BJ, Graham GH (1983) Dynamic stall regulation of the Darrieus turbine. Technical report, Sandia Report SAND83-7029 UC-261
21. Ouro P, Runge S, Luo Q, Stoesser T (2018) Three-dimensionality of the wake recovery behind a vertical axis turbine. Renew Energy
22. Kou J, Le Clainche S, Zhang W (2018) A reduced-order model for compressible flows with buffeting condition using higher order dynamic mode decomposition with a mode selection criterion. Phys Fluids 30(1):016103
23. Le Clainche S, Vega J (2017) Higher order dynamic mode decomposition. SIAM J Appl Dyn Syst 16(2):882–925
24. Le Clainche S, Vega J (2017) Higher order dynamic mode decomposition to identify and extrapolate flow patterns. Phys Fluids 29(8):084102

Suppressing Vortex Induced Vibrations of Wind Turbine Blades with Flaps

Sergio González Horcas, Mads Holst Aagaard Madsen, Niels Nørmark Sørensen and Frederik Zahle

Abstract The present work describes an exploratory work aiming to analyze the impact of trailing edge flaps activation on Vortex Induced Vibrations (VIV) suppression. A computational study of the VIV of the AVATAR rotor blade, a 10 MW design suitable for offshore locations, was performed. A Fluid Structure Interaction (FSI) approach was adopted for the simulations, coupling an Improved Delayed Detached Eddy Simulations (IDDES) flow solver with a beam-based structural model. Initial simulations based on the clean geometry identified significant edgewise VIV for certain free stream velocity and flow inclination angles. Inflow conditions showing the maximum amplitude of blade vibrations were used in order to test several trailing edge flap geometries and operating angles. The best flap configuration found in this parametric study managed to suppress the VIV phenomenon. However, when assessing a wider range of inflow conditions, the amplitudes of vibration of the blade equipped with flaps were found to be equivalent to the ones obtained for its clean counterpart. It is therefore concluded that a re-calibration of the flap operating angle should be required in order to adapt it to the considered wind speed and wind direction.

1 Introduction

Due to the flexibility of their blades, it is presumed that modern horizontal wind turbines may be subject to Vortex Induced Vibrations (VIV), a phenomenon which could ultimately lead to structural failure of the blade. This statement was supported by the recent high-fidelity simulations of [1], where the authors analyzed the DTU 10 MW reference wind turbine blade under realistic inflow conditions. The present work constitutes a first exploratory study in order to use flow control devices for the suppression of this phenomenon. In particular, the influence of the installation of trailing edge flaps is assessed. Previous studies based on Computational

S. González Horcas (✉) · M. H. A. Madsen · N. N. Sørensen · F. Zahle
DTU Wind Energy, Frederiksborgvej 399, 4000 Roskilde, Denmark
e-mail: sgho@dtu.dk

E. Ferrer and A. Montlaur (eds.), *Recent Advances in CFD for Wind and Tidal Offshore Turbines*, Springer Tracts in Mechanical Engineering,
https://doi.org/10.1007/978-3-030-11887-7_2

Fluid Dynamics (CFD) methods have shown the potential of these devices for load alleviation during the operation of multi-megawatt wind turbines [2, 3]. However, to the best of the authors' knowledge no publication regarding the use of trailing edge flaps for VIV suppression can be found in the literature.

In order to assess the performance of trailing edge flaps with regards to VIV suppression, a comprehensive numerical study was performed. It was based on the AVATAR reference rotor [4], that is equipped with a 100 m blade. The chosen computational method, presented in Sect. 2, relied on the coupling of a fluid and a structure solvers by means of a Fluid Structure Interaction (FSI) approach. Three different sets of FSI simulations were performed. First computations, detailed in Sect. 3, were based on the clean blade geometry. The objective of these simulations was to identify the specific inflow conditions which lead to the maximum amplification of vibrations, thereby highlighting the mechanisms involved in this phenomenon. A subsequent set of simulations accounting for trailing edge flaps was carried out. Due to time constraints, the deflected flaps geometry was estimated based on a simple analytical expression. The results of these simulations are presented in Sect. 4, and comprise both an assessment of the performance of the flap for VIV suppression as well as a quantification of the influence of the geometric parameters characterizing this device. For these simulations, the inflow conditions exhibiting the maximum amplification for the clean geometry were adopted. Finally, Sect. 5 compiles a wider assessment of the flaps installation with regards to VIV suppression. The initially studied analytical representation of the flaps was replaced by the use of a Free Form Deformation (FFD) method. This technology, introduced in the fluid solver in the framework of the present work, establishes a much more versatile approach for future applications. An initial comparison with the performance of the analytical flap version was performed, followed by the assessment of the inflow conditions range considered for the clean geometry.

2 Computational Set-Up

The fluid was modeled with the CFD solver EllipSys3D [5–7], which is a finite volume code that solves the Navier Stokes equations on a structured grid and in curvilinear coordinates. In all the simulations presented in this work, a zonal Improved Delayed Detached Eddy Simulation (IDDES) was employed in order to deal with turbulence [8]. In this way, a Reynolds-Averaged Navier Stokes (RANS) model was employed close to the blade surface and a Large Eddy Simulation (LES) approach was followed far away from the boundary layer (where the turbulent length scales become larger than the used grid resolution). For the region involving the RANS method, the k-ω SST turbulence model of [9] was used.

A three-level grid sequence was used during the simulation to speed up the development of the wake flow. A time step of $1.5 \cdot 10^{-3}$ s was employed, with a total simulation time of 68 s. The air density was fixed to 1.225 kg m^{-3}, and the dynamic viscosity was set to $1.7879 \cdot 10^{-5}$ kg m^{-1} s^{-1}. An inlet/outlet strategy was used for

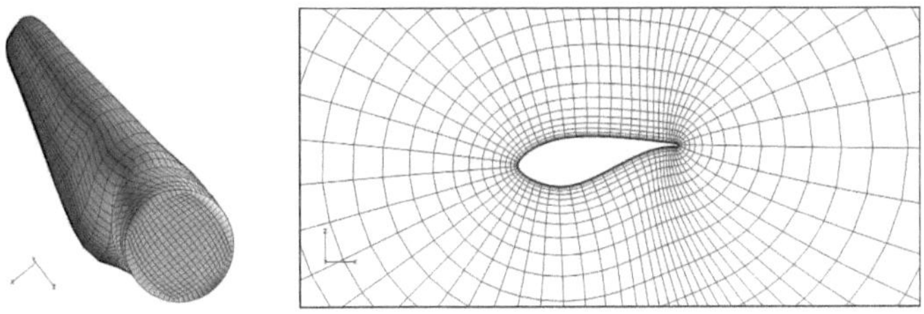

(a) Surface mesh (b) Half span cut

Fig. 1 Overview of CFD mesh. For clarity purposes, only 1 out of 4 grid lines are displayed

the boundary conditions. For the former region, a value of $1 \cdot 10^5$ s^{-1} was imposed for the specific dissipation ω, while the turbulent kinetic energy was set to $1 \cdot 10^{-2}$ m^2 s^{-2}.

The same structured CFD mesh was used for all simulations. It accounted for 240 blocks of 32×32×32 cells, leading to a total of 7.8 millions elements for the complete mesh. The chordwise direction was discretized with 256 cells (16 of them lying on the trailing edge). A boundary layer clustering was taken into account, with an imposed first cell height of around 10^{-6} m. This allowed to obtain y+ values lower than 1 for the considered simulation conditions. The surface mesh is depicted in Fig. 1, together with a detail of a section cut at half span.

The CFD model was coupled with a structural model of the blade by means of the staggered FSI approach presented by [1]. The inner CFD mesh deformation process was enhanced in the framework of the present work. A generalized analytical approach was implemented, where the deformation vector is smoothed based on a hyperbolic tangent function. The blade structure was handled by the commercial solver HAWC2 [10], and relied on a series of Timoshenko beam elements. The effect of both cone and tilt angles were neglected in the present study.

All the presented FSI computations were run on the `Jess` high-performance computing cluster owned by the Technical University of Denmark (DTU). A total of 240 processors running at 2.8 GHz were used per computation, requiring a wall clock time of approximately 15 h.

3 Results for the Clean Geometry

Previous computational studies of large wind turbines revealed that the main set-up parameters influencing the dynamic behavior of a standstill blade are related to the inflow conditions [1, 11]. For both aforementioned studies, maximum vibrations amplifications were observed for an angle of attack at the blade tip slightly higher than 90°. Hence, it was decided to fix this angle, referred to in this work as *pitch* angle θ, to 95°. Together with the pitch angle, both the absolute value of free stream velocity $\|U_\infty\|$ and its inclination angle with respect to the blade axis ψ were identified as

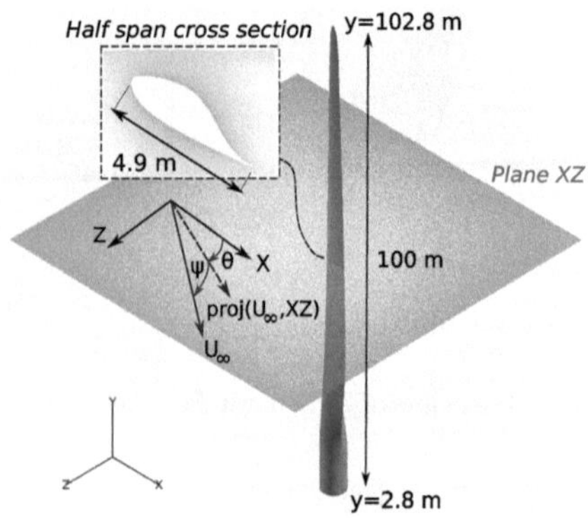

Fig. 2 Description of considered simulation parameters for the different test cases

the main actors triggering the VIV phenomenon. A graphical representation of these parameters is depicted in Fig. 2. All simulations included in this work are labeled as *UXX_IZZ*, where *XX* refers to the value of $\|U_\infty\|$ and *ZZ* to ψ.

In order to analyze the behavior of the clean blade geometry with regards to VIV, two consecutive parametric studies were performed. The first parametric study concerned the effect of the inclination angle ψ. In particular, angles ranging from 0 to 80° were considered. A common free stream velocity $\|U_\infty\|$ was used for all computations. It was estimated based on an analytical computation, aiming at the generation of outboard shedding frequencies at the vicinity of the first edgewise mode of the blade (i.e. 0.91 Hz), where maximum VIV are expected. Indeed, the natural frequency of shedding f_{shnat} of an inclined flow with an angle ψ can be related to the Strouhal number St by means of the independence principle, which is valid for small angles [12], as:

$$f_{shnat} = \frac{\|U_\infty\| \cos(\psi) St}{c} \tag{1}$$

where c refers to a reference length (in this case the chord). Assuming a Strouhal number of approximately 0.16 [1, 11], taking as reference chord the value at 85% of the blade span (i.e. 2.96 m) and assuming a complete synchronization of the shedding frequency and the first edgewise mode for an inclination angle of 40°, a $\|U_\infty\|$ of 22 m s^{-1} was selected for this initial study.

The second parametric study concerned the variation of the free stream velocity $\|U_\infty\|$. In particular, values ranging from 10 to 40 m s^{-1} were considered. The value of the inclination angle ψ was fixed to 40°, consistently with the assumptions made in the choice of the free stream velocity of the first parametric study.

Figure 3 depicts, for both parametric studies, the maximum observed peak-to-peak values of the blade tip deflection transients. Since the VIV phenomenon is expected for the edgewise motion, only the tangential component with respect to the rotor plane is shown (corresponding in this case to the x-direction). Regarding the parametric study of the inclination angle (Fig. 3a), the results reveal a significant amplification of the edgewise deflection for the U22_I40 simulation, with a maximum

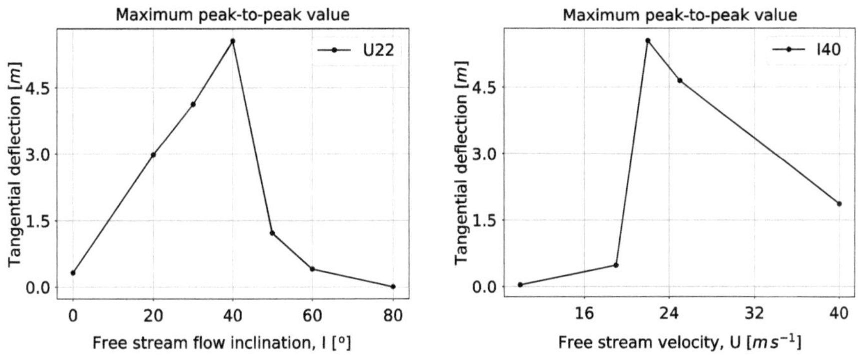

(a) Effect of inclination angle

(b) Effect of free stream velocity

Fig. 3 Maximum peak-to-peak value of the tangential blade tip deflection transients. Clean blade

peak-to-peak value of 5.55 m (corresponding to 9.25 times the blade tip chord). The variation of the inclination angle led to a decrease of the computed deflections, with negligible amplitudes for both 0 and 80°. An analogous observation can be made for the parametric study concerning the influence of the free stream velocity, shown in Fig. 3b. Due to the desynchronization of the shedding and the first edgewise frequencies, a decrease in the amplitude of deflections was observed when assessing wind speed values deferring from 22 $m\,s^{-1}$.

In Fig. 4, the Power Spectral Density (PSD) of the blade sectional loading time series is shown for several simulations. Due to its potential impact on edgewise vibrations, only the x-component of the load was considered. Additionally, the plots were limited to a high span range. Results are displayed by means of color plots, with the ordinates representing the blade radial position and the corresponding load frequencies included in the abcissae. All sectional PSDs were computed by means of the Welch method [13], and the results were linearly scaled (so that the peaks height become an estimate of the RMS amplitude). For the U22_I40 simulation (Fig. 4a), which exhibited the highest vibration amplitudes, important loads fluctuations were observed. Additionally, a significant correlation of the loading frequencies along the outboard region of the blade was found, corresponding to the first edgewise mode of the blade. This fact, already observed in [1, 11], is assumed to be related to the the existence of the spanwise flow. Indeed, the maximum fluctuations were obtained at 85% of the blade span, as initially predicted by Eq. 1, corresponding to y = 87.8 m. This significant fluctuation was then convected all along the span, resulting in a high frequency content around the first edgewise mode for the outboard region of the blade. Much lower PSD levels were obtained for the U22_I0 simulation (Fig. 4b), were the absence of a spanwise flow component led to the decorrelation of the loading frequencies. Finally, the results of the U10_40 simulation are depicted in Fig. 4c. For this simulation, a considerable difference between the natural shedding frequencies of the flow and the first edgewise mode is expected, due to the low value of the free stream velocity. This could explain the low PSD values computed for this simulation.

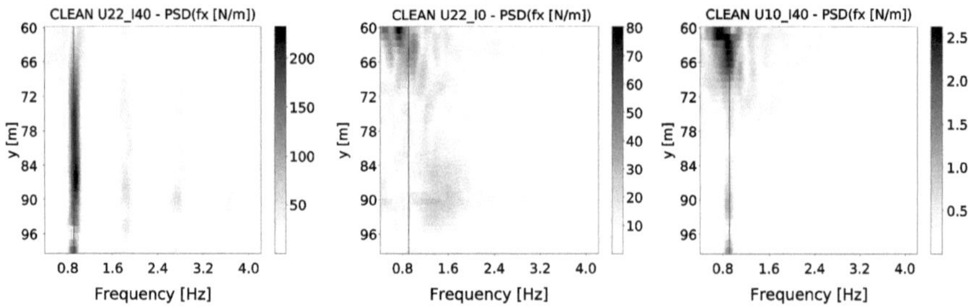

(a) U22_I40 (b) U22_I0 (c) U10_I40

Fig. 4 PSDs of the x-component of blade sectional loading for the clean blade. A vertical line corresponding to the first edgewise mode of the blade was added for reference. Units are $N^2\,m^{-2}$

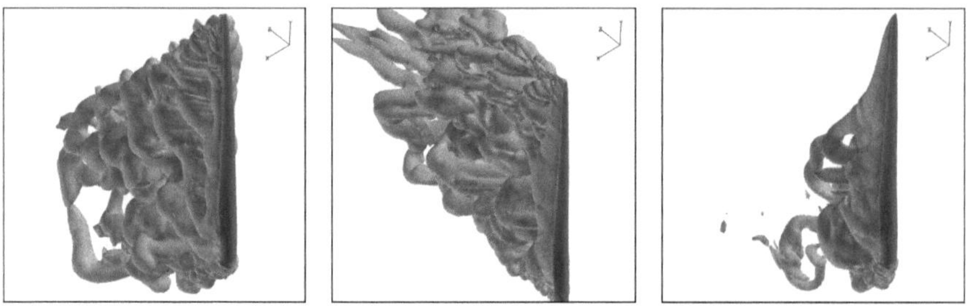
(a) U22_I40 (b) U22_I0 (c) U10_I40

Fig. 5 Iso-surface of absolute value of vorticity at $2\,s^{-1}$. Different inflow conditions for clean blade

Indeed, a suppression of the Von Karman street for the outboard region of the blade was observed. This is illustrated in Fig. 5 by means of the iso-surfaces of the absolute value of vorticity, where analogous results for the two other simulations are included for reference.

4 Preliminary Study Based on an Analytic Flap

In order to assess the performance of the introduction of flaps on the suppression of VIV, a preliminary study based on an analytical hinged flap definition was performed. Only the inflow conditions showing the maximum vibration amplitudes for the clean configuration, i.e. U22 and I40, were considered in this study. The flap deflection was introduced by means of a local deformation of the CFD mesh of the blade surface.

Two main geometrical parameters were considered: the flap length with respect to the chord and the spanwise extent with respect to the blade radius. In particular, values of 20% and 30% were considered for the former variable, labeled in this work as *XC20* and *XC30*. Regarding the spanwise extent, values of 20% and 25% were considered, referred to in this work as *L020* and *L025*. For both cases, the flap was

Fig. 6 Geometry of the analytic flap XC20_L20 operating at $-30°$

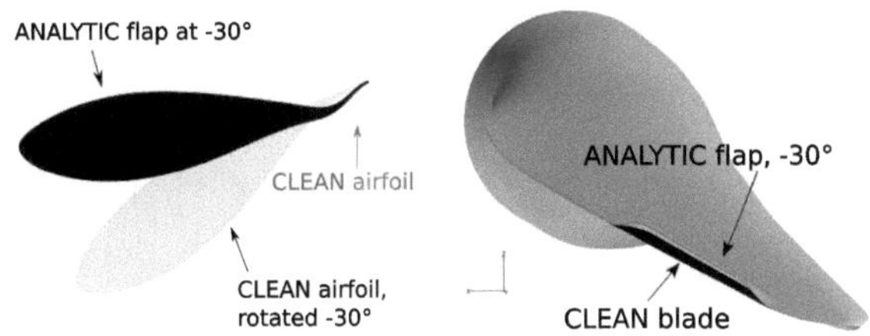

(a) Half of flap extent (b) 3D view

assumed to be centered at 75% of the blade radius. This led to a total radial extent ranging from y = 68 m to y = 88 m for the *L020* configuration, and from y = 65 m to y = 90 m for the *L025* case. Four different operating angles were also considered: -30, 10, 10 and 30°. Those are labelled in this work as *DN30*, *DN10*, *D10* and *D30* respectively.

The deflected flap geometry was achieved by a linear interpolation of the clean mesh and one rotated according to the flapping angle (see Fig. 6a). A fifth order polynomial smoothing function was applied from the position of the hinge point, estimated based on the value of the flap length, up to the trailing edge. This ensured a smooth transition of the suction and the pressure side curvatures, even if a higher effective flapping angle was expected. An analogous smoothing function was applied in the spanwise direction in order to ensure a smooth transition between the flapped and clean configurations (see Fig. 6b).

Figure 7 compiles the blade tip tangential deflection transients for the studied flapping geometries, superposed to the corresponding results for the clean configuration. From the four studied flapping angles (i.e. $-30°$, $-10°$, 10° and 30°) only those leading to a significant reduction of blade tip deflections are shown. In particular for $-30°$, all configurations managed to suppress the amplification of the blade vibration, regardless of the flap geometry (see Fig. 7b). The same remark can be made for the simulations accounting for a flap extent of 25% in the spanwise direction and operating at 10°(dashed lines in Fig. 7a). For both cases, the mechanism preventing the amplification of the blade vibration was assumed to be related to the suppression of the Von Karman street at the outboard region. This is illustrated in Fig. 8 by means of the iso-surfaces of vorticity of a flap configuration undergoing VIV (XC20_L20_D30), and two set-up where the vibrations were suppressed (XC20_L20_DN30 and XC20_L25_D10). A more detailed visualization of the Von Karman street suppression is depicted in Fig. 9, where the y-component of the vorticity is shown for two different blade sections. All simulations revealed a spanwise vortex generated at the blade tip (see Fig. 9a, b, c). For XC20_L20_DN30 and XC20_L25_D10 computations, this spanwise vortex remained attached up to approximately 65% of the blade span (see Fig. 9e, f). For lower sections, all studied computations exhibited a Von Karman street similar to the one found for XC20_L20_D30, depicted in Fig. 9d.

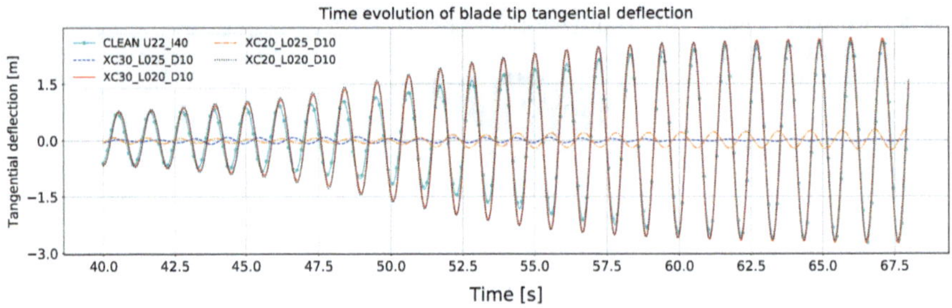

(a) Flap angle 10°

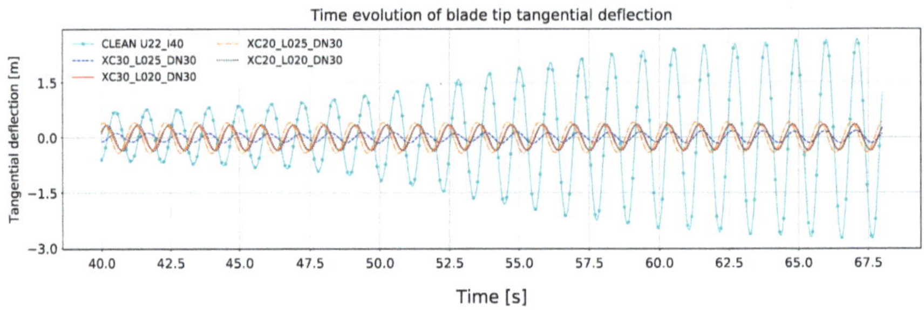

(b) Flap angle -30°

Fig. 7 Blade tip tangential deflection transient for different analytic flap configurations and operating angles

(a) XC20_L20_D30 (b) XC20_L20_DN30 (c) XC20_L25_D10

Fig. 8 Vorticity iso-surface at 2 s^{-1}, blade equipped with different analytic flap geometries. Flapping angles set to 30°(a), −30°(b) and 10°(c)

The detailed mechanism preventing the outboard region Von Karman street would require additional work, including a pure CFD study of a stiff blade configuration. However, the change of the Strouhal number induced by the flap geometry could be argued as a preliminary explanation. This is consistent with the reduction of VIV observed for the clean configuration at low wind speeds (see Sect. 3), where significant differences between the natural shedding frequencies of the blade and the first edgewise mode were expected. To the best of the authors' knowledge, this is the first time that this complex three-dimensional flow behaviour is identified in the context of wind energy. However, an extensive literature dealing with inclined

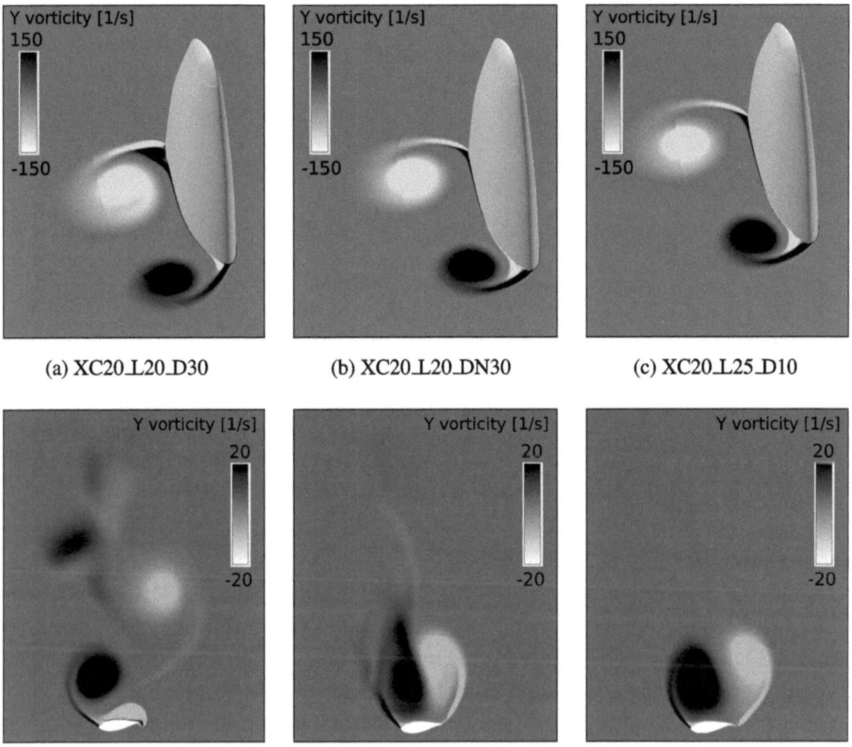

(a) XC20_L20_D30 (b) XC20_L20_DN30 (c) XC20_L25_D10

(d) XC20_L20_D30 (e) XC20_L20_DN30 (f) XC20_L25_D10

Fig. 9 Y vorticity component for section cuts located near the blade tip (upper row) and at y = 77.9 m (lower row), for blade equipped with different analytic flap geometries

cylinders can be found. The existence of the spanwise vortex was identified, among others, in the experimental work of [14]. The mitigation of VIV by the suppression of the Von Karman Street was reported in the DES simulations of [15]. This was achieved by equipping an inclined cylinder with strakes. Similar conclusions were found in the subsequent computational studies of [16, 17].

The suppression of the Von Karman street led to a mitigation of the load fluctuations, explaining the low level of observed blade vibrations. This is illustrated in the PSDs of the blade sectional loading of Fig. 10. It can be observed that for the blade equipped with a not optimal flap configuration (Fig. 10a), the computed load fluctuations were similar to those obtained for the clean blade, previously shown in Fig. 4a. For the other two flap configurations (Fig. 10b, c), the energy content around the first edgewise was significantly decreased.

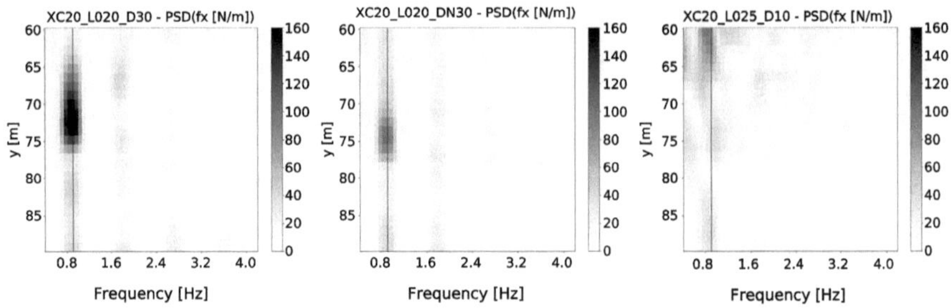

(a) XC20_L20_D30 (b) XC20_L20_DN30 (c) XC20_L25_D10

Fig. 10 PSDs of the x component of blade sectional loading. Includes several analytic flap configurations. A vertical line corresponding to the first edgewise mode of the blade was added for reference

5 Introduction of the Free form Deformation Flap

The preliminary study based on an analytical definition of the flap revealed a good potential for all studied geometries when working at −30°. The XC20_L20 was therefore selected, due to its reduced dimensions, to be re-evaluated by means of the Free Deformation (FFD) method. This technique, originating from the field of computer graphics [18], is a very lenient and versatile way of manipulating a given geometry. The most straightforward way of understanding the deformation mechanism is to think of the object as being wrapped in soft rubber. When one moves the rubber material (by moving given control points) the deformation is propagated inwards to deform all material inside the control box depending on the distance to the moved control points. There are many benefits of FFD including possible continuity control, volume preservation and analytically defined deformation gradients. The FFD utilities used in the present work are all from the `FFDlib`-toolbox, which is developed at DTU Wind Energy to facilitate High-Fidelity Shape Optimization as well as general surface mesh deformation. In Fig. 11a a box from `FFDlib` is shown along the trailing edge of the blade, which is displayed in black. By moving only a selection of control points (shown in white), a smooth transition from flapping region to original blade is obtained. Figure 11 shows a comparison of the analytic and FFD flap versions. It should be remarked that the FFD implementation accounts for a faster transition towards the rotated flap geometry. This results in a lower (and more realistic) flapping angle. In particular, an analytic flap working at −30° seemed to correspond to the FFD counterpart operating at −50°.

First simulations were devoted to verify this observation under the U22_I40 inflow conditions. Figure 12 shows a comparison of the FFD flap working at several negative flapping angles. As expected, a good agreement was found for the FFD flap working at −50°. This operation angle was kept for the final evaluation of the flap, where the whole set of inflow conditions presented in Sect. 3 was considered.

Figure 13 depicts the maximum peak-to-peak values computed for the blade equipped with the FFD flap, compared to the clean geometry results. Outside the

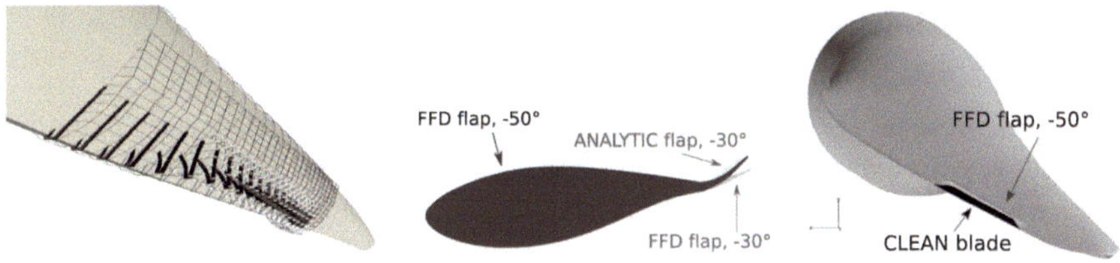

(a) FFD box (b) Half of flap extent (c) 3D view

Fig. 11 Geometry of the FFD flap XC20_L20 operating at $-50°$

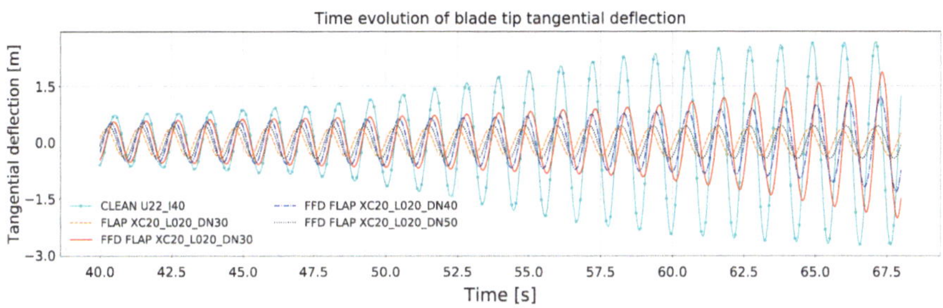

Fig. 12 Blade tip tangential deflection transient for the analytic and FFD flaps. U22_I40 simulation

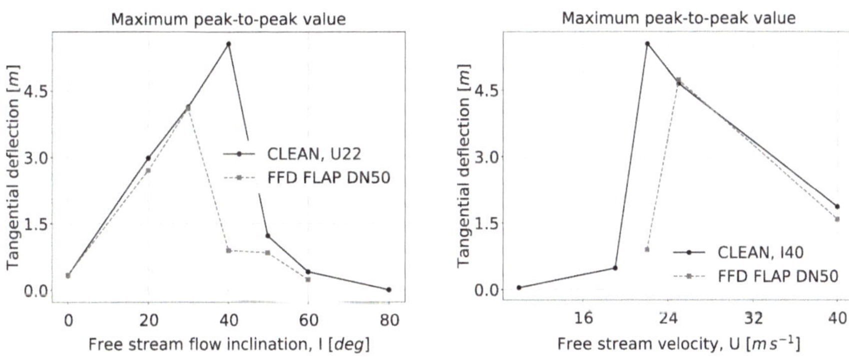

(a) Effect of inclination angle (b) Effect of free stream velocity

Fig. 13 Maximum peak-to-peak value of the tangential blade tip deflection transients. Comparison of clean geometry and FFD flap working at $-50°$

inflow conditions showing the maximum blade vibrations amplification, the behavior of the flapped blade was found to be equivalent to its clean counterpart. This reveals that the choice of the flapping angle performed in Sect. 4 should be re-assessed based on the considered free stream velocity vector.

6 Conclusion

This work presents a computational study aiming to assess the performance of trailing edge flaps with regards to the suppression of Vortex Induced Vibrations (VIV) of large wind turbine blades. In particular the blade of the the AVATAR reference rotor was studied by means of a Fluid Structure Interaction (FSI) approach, coupling an Improved Delayed Detached Eddy Simulation (IDDES) solver with a beam model of the structure.

A first set of computations based on the clean blade geometry allowed to determine several inflow configurations triggering the VIV phenomenon. As for previous work included in the literature, two main parameters leading to the amplification of the blade edgewise vibrations were identified. On the one hand, the chosen absolute value of the free stream velocity should be tuned in order to lead to natural shedding frequencies at the outboard region close to the first edgewise mode of the blade. As shown in this paper, a good initial estimation can be made by assuming a typical Strouhal number of 0.16. The second parameter required for the triggering of VIV was found to be the inclination angle of the flow with respect to the blade axis. Indeed, a significant spanwise correlation of the shedding frequencies of the different blade sections was achieved by considering inclination angles of around 40°.

A second set of simulations accounted for trailing edge flaps, defined *via* an analytical expression of their geometry. The inflow conditions exhibiting the maximum amplification for the clean geometry were assumed. The performance of the flaps for VIV suppression was assessed, together with the influence of the geometric parameters characterizing this device. Several flap geometries did manage to suppress the blade vibrations amplification. Of particular interest was the operating angle −30°, that resulted in very low blade tip deflection amplitudes regardless of the considered flap geometry. The mechanism preventing the VIV was concluded to be related to the suppression of the Von Karman street at the outboard region. Further studies would be required in order to completely understand this phenomenon. As a preliminary hypothesis, the change of the Strouhal number distribution in the outboard region induced by the flap geometry was argued.

The final set of simulations aimed to evaluate the performance of trailing edge flaps for VIV suppression in a wider range of inflow conditions. The initially studied analytical representation of the flap geometry was replaced by the use of a Free Form Deformation, allowing a much more versatile approach for future applications. A single flap operating angle, calibrated in previous simulations, was also used for all the new inflow conditions. The obtained results showed that, outside of the initially studied inflow conditions, the behaviour of the flapped geometry was equivalent to its clean counterpart. It was therefore concluded that a re-calibration of the flap operating angle should be required in order to adapt it to the inflow conditions. In view of the parallelism between the observed Von Karman street suppression and previous experiences regarding straked cylinders, the performance of the studied flaps could also be improved by considering variable operating angles in spanwise.

This first evaluation of the use of flaps for VIV suppression reveals a good potential for these flow control devices. However, a better understanding of the involved FSI mechanisms should be achieved. This would allow for an identification of the best operating angle for every inflow configuration, together with a more realistic design of the flap geometries. In particular, stiff simulations of the flapped geometries would be required in order to evaluate the change in the Von Karman shedding patterns. The introduction of the FFD approach in the proposed numerical approach also allows for an easier exploration of new flow control devices as a future work, such as the so-called split flaps or a combination of trailing edge flaps working at different operating angles.

References

1. Heinz JC, Sørensen NN, Zahle F, Skrypinski W (2016) Vortex-induced vibrations on a modern wind turbine blade. Wind Energy 19(11):2041–2051
2. Barlas T, Jost E, Pirrung G, Tsiantas T, Riziotis V, Navalkar ST, Lutz T, Van Wingerden JW (2016) Benchmarking aerodynamic prediction of unsteady rotor aerodynamics of active flaps on wind turbine blades using ranging fidelity tools. J Phys: Conf Ser 753(2)
3. Jost E, Fischer A, Lutz T, Krämer E (2016) An investigation of unsteady 3D effects on trailing edge flaps. J Phys: Conf Ser 753(2)
4. Lekou D, Chortis D, Chaviaropoulos P, Munduate X, Irisarri A, Madsen HA, Yde K, Thomsen K, Stettner M, Reijerkerk M, Grasso F, Savenije R, Schepers G, Andersen C (2015) AVATAR Deliverable D1.2: reference blade design. Technical report, ECN Wind energy, Petten, The Netherlands
5. Michelsen JA (1992) Basis3D - A platform for development of multiblock PDE solvers. Technical report AFM 92-05, Department of Fluid Mechanics, Technical University of Denmark
6. Michelsen JA (1994) Block structured multigrid solution of 2D and 3D elliptic PDE's. Technical report AFM 94-06, Department of Fluid Mechanics, Technical University of Denmark
7. Sørensen NN (1995) General purpose flow solver applied to flow over hills. Risø-R- 827-(EN), Risø National Laboratory, Roskilde, Denmark
8. Menter FR, Kuntz M (2004) Adaptation of eddy-viscosity turbulence models to unsteady separated flow behind vehicles. In: McCallen R, Browand F, Ross J (eds) The aerodynamics of heavy vehicles: trucks, buses, and trains. Springer, Berlin, pp 339–352
9. Menter FR (1994) Two-equation eddy-viscosity turbulence models for engineering applications. AIAA J 32(8):1598–1605
10. Larsen TJ, Hansen AM (2015) HAWC2, the user's manual. Technical report July, Risø
11. Heinz JC, Sørensen NN, Riziotis V, Schwarz M, Gomez-iradi S, Stettner M (2016) Aerodynamics of large rotors. WP4. Deliverable 4.5. Technical report, ECN Wind Energy, Petten, The Netherlands
12. Hoang MC, Laneville A, Légeron F (2015) Experimental study on aerodynamic coefficients of yawed cylinders. J Fluids Struct 54:597–611
13. Welch P (1967) The use of the fast Fourier transform for the estimation of power spectra: a method based on time averaging over short, modified periodograms. IEEE Trans Audio Electroacoust 15:70–73
14. Matsumoto M, Yagi T, Shigemura Y, Tsushima D (2001) Vortex-induced cable vibration of cable-stayed bridges at high reduced wind velocity. J Wind Eng Ind Aerodyn 89(7–8):633–647
15. Yeo D, Jones NP (2010) Aerodynamic effects of strake patterns on flow around a yawed circular cylinder. In: The fifth international symposium on computational wind engineering (CWE2010), Chapel Hill, North Carolina, USA
16. Gioria RS, Korkischko I, Meneghini JR (2011) Simulation of flow around a circular cylinder fitted with strakes. In: 21st International congress of mechanical engineering

17. Zhou T, Razali SF, Hao Z, Cheng L (2011) On the study of vortex-induced vibration of a cylinder with helical strakes. J Fluids Struct 27(7):903–917
18. Sederberg TW, Parry SR (1986) Free-form deformation of solid geometric models. ACM SIGGRAPH Comput Graph 20(4):151–160

Prediction of the Wake Behind a Horizontal Axis Tidal Turbine Using a LES-ALM

Pablo Ouro, Magnus Harrold, Luis Ramirez and Thorsten Stoesser

Abstract A large-eddy simulation-actuator line method (LES-ALM) applied to a single horizontal axis tidal turbine is presented and validated against experimental data. At a reasonable computational cost, the LES-ALM is capable of capturing the complex wake dynamics, such as tip vortices, despite not explicitly resolving the turbine's geometry. The LES-ALM is employed to replicate the wake behind a laboratory-scale horizontal axis turbine and achieves a reasonably good agreement with measured data in terms of streamwise velocities and turbulence intensity. The turbine is simulated at six tip speed ratios in order to investigate the rate of decay of velocity deficit and turbulent kinetic energy. In the far-wake, these quantities follow a similar decay rate as proposed in the literature with a −3/4 slope. For cases when the turbine spins at or above the optimal tip speed ratio, the levels of turbulent kinetic energy and wake deficit in the far-wake are found to converge to similar values which seem to be linearly correlated. Finally, transverse velocity profiles from the simulations agree well with those from an analytical model suggesting that the LES-ALM is well-suited for the simulation of the wake of tidal stream turbines.

P. Ouro (✉)
School of Engineering, Hydro-environmental Research Centre, Cardiff University, CF24 3AA Cardiff, United Kingdom
e-mail: ourobarbap@cardiff.ac.uk

M. Harrold
College of Engineering, Mathematics and Physical Sciences, University of Exeter, TR10 9FE Penryn, United Kingdom
e-mail: m.j.harrold@exeter.ac.uk

L. Ramirez
Group in Numerical Methods in Engineering, University of A Coruña, Campus de Elviña, 15071 A Coruña, Spain
e-mail: luis.ramirez@udc.es

T. Stoesser
Civil, Environmental and Geomatic Engineering, University College London, WC1E 6BT London, United Kingdom
e-mail: stoesser@ucl.ac.uk

E. Ferrer and A. Montlaur (eds.), *Recent Advances in CFD for Wind and Tidal Offshore Turbines*, Springer Tracts in Mechanical Engineering,
https://doi.org/10.1007/978-3-030-11887-7_3

1 Introduction

Most tidal energy projects aim at deploying several devices in close proximity to each other to maximise the power output and to minimise the logistical challenges at selected offshore deployment sites. Tidal turbines harness the kinetic energy by extracting momentum from the fast-flowing tidal streams and, as a consequence, generate a low-momentum wake behind them [1]. Turbines operating in the wake of turbines which are located upstream face the challenges of higher turbulence levels and lower approach flow velocity [2]. Thus, the design of turbine farms requires knowledge of the turbine-turbine inter-play in order to avoid sub-optimal performance and high hydrodynamic loads when turbines are situated in the wake of others.

In this context, accurate numerical methods are needed to confidently study different array configurations under varying flow conditions [3]. Eddy-resolving methods, such as Large-Eddy Simulation (LES), are thus required to accurately resolve the highly turbulent nature of the turbines' wakes [1] because time-averaged approaches, such as Reynolds-Averaged Navier Stokes (RANS) models, are unable to capture the large-scale unsteadiness in the flow around tidal turbines.

To date, investigations of tidal stream turbines have focused primarily on the performance of individual devices operating at their peak efficiency as it is the ideal scenario. However, the experience gained in offshore wind energy suggests that once the turbines are deployed in multiple rows, wake-shadowing effects arise [4]. This suggests that turbines in deployed in rows, except for the first one, may operate in lower approach flow velocities as predicted. Hence, axial induction control or wake-steering via yaw-angle adjustment are strategies that aim to maximise the farm's performance as a whole instead of maximising the output of a few individual turbines [5].

To expand the current understanding of wakes behind tidal stream turbines at various rotational speeds, this paper presents an LES code equipped with an actuator line model to be validated with experimental data. The method is then used to predict the decay of velocity deficit and turbulence in a tidal turbine's wake.

2 Computational Model

Simulations are performed using the in-house large-eddy simulation code Hydro3D [6], which solves the spatially filtered Navier-Stokes equations for incompressible viscous flow. Hydro3D discretises the computational domain using rectangular Cartesian grids with staggered storage of velocities, i.e. the pressure values are stored in the cell centres whilst the three components of velocity are in the cell faces. The computational domain can be decomposed into smaller sub-domains via Domain Decomposition in order to compute the simulation in parallel using Message Passing Interface (MPI). A fractional-step method is adopted to advance the simulation in time together with a 3-step Runge–Kutta algorithm. An efficient multi-grid technique is used for calculating the Poisson pressure equation that is normally the most expensive operation in LES computations for incompressible flows, and the WALE sub-grid scale model is adopted to approximate the sub-grid stress tensor τ. Details of the LES code can be found in [7, 8].

Here an Actuator Line Model (ALM) is implemented in the LES solver in order to represent horizontal axis tidal turbines at an efficient computational cost. Ouro et al. [1] developed an Immersed Boundary Method (IBM) in Hydro3D to perform geometry-resolved simulations of a tidal turbine, achieving good accuracy in the prediction of the hydrodynamic forces and turbulent wake. The computational expense of using the IBM is relatively high and for an entire array of turbines it would require thousands of CPUs, whilst the ALM arises as a promising alternative because it provides a good balance between accuracy and computational expense.

The ALM divides each turbine blade into a number of sections of length Δr of given chord length, c, and prescribed hydrodynamic characteristics, i.e. tabulated lift and drag coefficients (C_L, C_D). Each section is represented by a Lagrangian marker that moves freely along the computational mesh and delta functions are responsible for transferring velocities and forces between Lagrangian and Eulerian frameworks [9]. At every time step, the relative velocity (V_{rel}) at each marker is calculated in order to determine the Lift (L) and Drag (D) forces as,

$$\mathrm{L} = \frac{1}{2}\rho|V_{rel}|^2 C_L \cdot \Delta A \cdot \mathrm{F1} \tag{1}$$

$$\mathrm{D} = \frac{1}{2}\rho|V_{rel}|^2 C_D \cdot \Delta A \cdot \mathrm{F1} \tag{2}$$

where $|V_{rel}|$ is the module of the relative velocity vector, $\Delta A = c\Delta r$ is the area associated to each marker, and F1 is a tip-loss correction term [10] that reads,

$$\mathrm{F1} = \frac{2}{\pi}\mathrm{acos}\left(\exp\left(-g\frac{N_b(R-r)}{2r\sin\phi}\right)\right) \tag{3}$$

here g is a function that depends on the number of blades N_b comprising the turbine and Tip Speed Ratio (TSR, $\lambda = \Omega R/U_0$, where Ω is the turbine's rotational speed, R is the turbine's radius and U_0 is the free-stream velocity), and is calculated as,

$$g = \exp(-0.125(N_b\lambda - 21)) + 0.1 \tag{4}$$

Thrust and torque are calculated by the projection of the lift and drag forces onto the flow direction and plane of rotation, respectively. Coefficients of thrust (C_x) and power (C_p) from the entire rotor are determined as,

$$C_x = \frac{F_x}{\frac{1}{2}\rho U_0^2 \pi R^2} \quad ; \quad C_p = \frac{Q\Omega}{\frac{1}{2}\rho U_0^3 \pi R^2} \tag{5}$$

where F_x represents the forces in the direction of the flow acting on the turbine's rotor and Q is the torque from all the turbine blades. Note that the turbines' hub and vertical support structures are modelled using the IBM.

2.1 Experimental Test and Computational Setup

Stallard et al. [2] tested a Horizontal Axis Tidal Turbine (HATT) in a hydraulic flume measuring 11.5 m long, 5.4 m wide and 0.45 m deep (h). The inflow velocity profile was measured along most of the water column and followed a power law distribution with a mean value (U_0) of 0.47 m/s. A turbulence intensity in the flow direction ($I_x = u'/U_0$, with u' as the root-mean-square velocity fluctuation) of 12% was found with turbulence length scales $L_x = 0.56$h, $L_y = 0.33$h and $L_z = 0.55$h. The scaled prototype had a diameter $D = 0.27$ m with blades designed with Goettingen 804 airfoil sections and the rotor rotated at a fixed tip speed ratio of $\lambda = 4.5$ coinciding with the operation point of maximum performance. The blades were attached to a 0.05 m long circular hub subsequently connected to a vertical tower that united the rotor with the external support structure.

The computational domain is geometrically identical to the flume. The grid resolution is uniform in the three spatial directions. Two meshes are employed, a coarse grid where $\Delta x = 0.010$m and a fine grid where $\Delta x = 0.005$ m in the local mesh refinement region in which the turbine is embedded. These mesh resolutions lead to 27 and 54 elements along the rotor's diameter, i.e. $\Delta x/D = 0.035$ and 0.0185 respectively, which are commonly adopted to represent wind turbines [11]. Fixed time steps (Δt) are used with values of 0.002 s and 0.001 s, and these are tested on both coarse and fine grids during the mesh sensitivity study in Sect. 3. The experimental velocity profile is prescribed at the inlet and the synthetic eddy method superimposes turbulence fluctuations onto the mean velocity with the integral turbulence length scales identical to those measured experimentally. Wall functions are imposed at the bottom and lateral walls, a convective outlet condition is set at the outlet and a shear-free slip condition is used at the top.

3 Results

3.1 Predicted Hydrodynamic Coefficients

The present LES-ALM is first validated in terms of predicted thrust and power coefficients at six different tip speed ratios in comparison to experimental data [2] and RANS-BEM results [12], and these are presented in Fig. 1. The turbine's peak performance, i.e. maximum C_p, is attained at $\lambda = 4.5$ and it is well-predicted by the LES-ALM, although with a value of $C_P = 0.33$ slightly over-predicting the experimental value of $C_P = 0.27$. A closer agreement with experiments in terms of C_p is found for the other rotational speeds. The blue line represents RANS predictions of C_P and these do not match as well with the experiments as the LES-ALM predictions.

At $\lambda = 4.5$, the thrust coefficient obtained with the LES-ALM is $C_x = 0.90$ that is slightly superior to $C_x = 0.85$ measured in the experiments. Overall, the LES-ALM results are in a reasonably good agreement with the experimental data

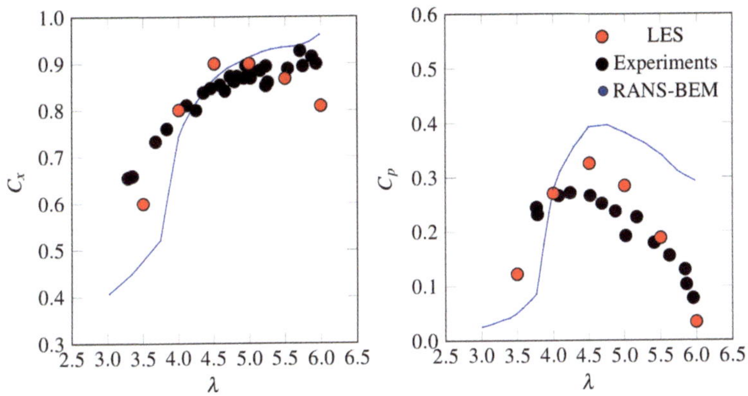

Fig. 1 Coefficients of trust (C_x) and power (C_p) from experiments [2], RANS-BEM [12] and present LES results obtained for different tip speed ratios

considering that the ALM does not explicitly resolve the turbine's geometry and relies on the parametrisation of most of the turbine's fluid-structure-interaction, such as flow separation from the blades or dynamic stall, only through tabulated lift and drag coefficients.

3.2 Mean Flow Field

Sensitivity of the LES-ALM to the mesh resolution and time step is analysed with the help of Fig. 2 which presents transverse profiles of velocity deficit ($\Delta U = U - U_0$) and turbulence intensity (I_x) at locations 2, 4, 6, 8 and 12D downstream of the turbine rotating at its optimum performance. Results of ΔU for the coarse mesh ($\Delta x = 0.02$ m) at x/D = 2 and 4 demonstrate that the model does not capture the peak of the momentum deficit. This is improved when using a finer mesh and smaller time step although, at x/D > 6, the LES-ALM still overestimates the velocity deficit in the centreline, and a minimal variation in the results is obtained regardless of the time step used.

At the location furthest downstream, i.e. x/D = 12, the three numerical configurations show a similar velocity distribution, underestimating the deficit recovery in the wake's centre and its lateral expansion. Park et al. [13] discussed that using synthetic turbulent inflow conditions can influence the far-wake mixing predictions as turbulence properties, e.g. velocity fluctuations and shear stresses, are interrelated and imposing them independently can distort the turbulence characteristics differing from those desired.

Profiles of turbulence intensity show that those set-ups in which the ratio of $\Delta x/\Delta t$ is equal to 10 captured the two-peak distribution of I_x at x/D = 2, and how it becomes more uniformly distributed in the lateral direction, as nearly accomplished at x/D = 12D with a value of $I_x \approx 0.1$. On the other hand, the simulation using a

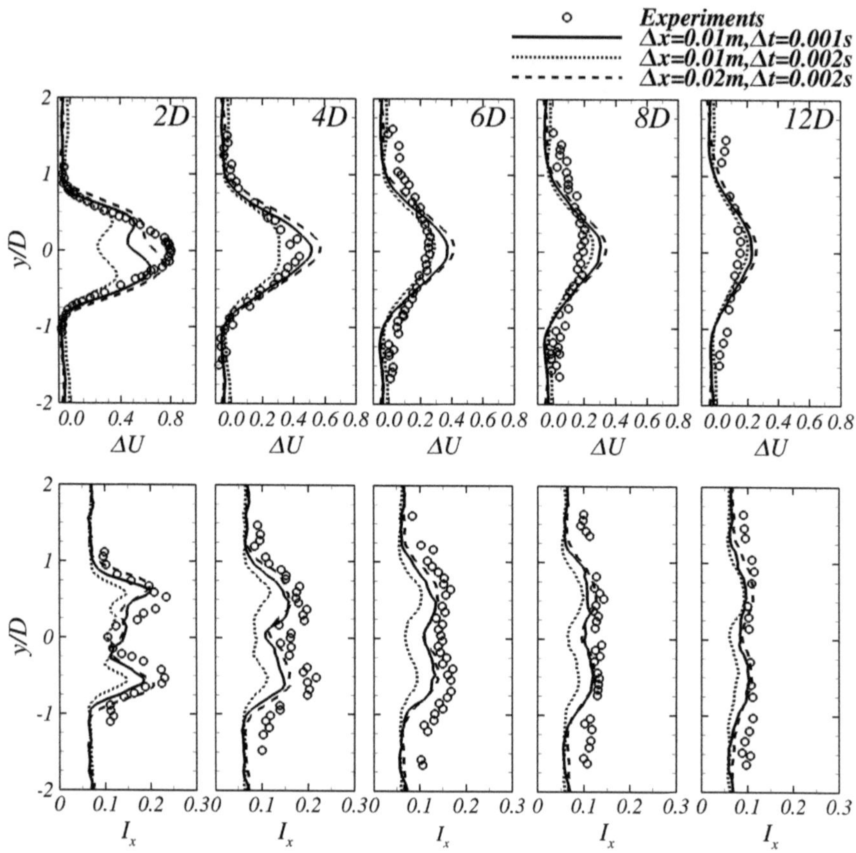

Fig. 2 Transverse profiles of velocity deficit (top row) and streamwise turbulence intensity (bottom row) at five different locations downstream of the turbine

time step of 0.002 s and the fine grid leads to an under-prediction of the levels of turbulence intensity throughout the wake. These results suggest that to obtain an accurate prediction of the wake dynamics there needs to be a specific ratio between the grid resolution and time step in the LES-ALM. If such proportion is ensured, the LES-ALM predicts the flow statistics with similar accuracy independently the fine or coarse mesh is adopted.

The flow field developed behind the simulated tidal stream turbine rotating at $\lambda = 4.5$ is presented in Fig. 3 with contours of velocity deficit and turbulence intensity. The region of greatest velocity deficit is found immediately behind the turbine due to the interaction of the moving blades, hub and support structure with the flow. Despite momentum being recovered at a few diameters downstream of the turbine, its signature in the mean flow field is still visible until 20–24D downstream.

Figure 3b depicts two regions of large turbulence intensity right behind the device. The first is found in the path followed by the tip vortices due to the shear induced by the moving blades on the oncoming flow [1], which is observed until

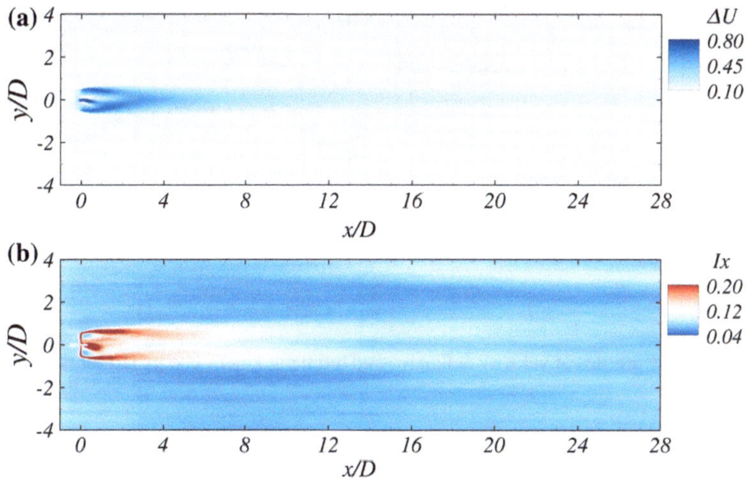

Fig. 3 Contours of (**a**) mean velocity deficit and (**b**) turbulence intensity at a horizontal plane at hub height

x/D $\approx$ 4, since this is where these energetic flow structures lose coherence. The vertical cylinder supporting the turbine also generates flow separation as observed in an area of $I_x > 0.15$ found at y/D = 0 until x/D = 1.

3.3 Centreline Wake Recovery

The downstream evolution of velocity deficit and turbulent kinetic energy ($k/U_0^2 = 0.5u_i'u_i'$, with $i = 1,2,3$) along the centreline of the wake is presented in Fig. 4 for the range of tip speed ratios simulated. At all rotational speed, a minimum in the velocity deficit distribution is attained at x/D $\approx$ 1 (represented as filled ∘ symbols) due to the flow going through the rotor's swept perimeter not being fully blocked, as seen in Fig. 3a [14]. Thereafter, the velocity deficit increases again until a second maximum is achieved, i.e. the mean streamwise velocity is minimum (represented with filled △ symbols). This transition stage in the wake occurs at a location (X_t) which is closer to the turbine with increasing rotational speed. The value of the ΔU, which peaks at X_t, also increases when the turbine operates at a faster rotational speed, indicating it generates a larger obstruction to the flow.

The profiles of the turbulent kinetic energy for all TSR peak at x/D $\approx$ 0.5 as a result of the interaction between the wake induced by the rotating blades and the flow separation in the lee side of the support structure. The peaks have a similar value of $k/U_0^2 \approx 0.08$ for $\lambda > 4.5$ and are notably reduced for $\lambda = 3.5$ and 4.0. Thereafter, k decays with increasing distance along the wake centreline. For these cases, the turbine operates over its optimal tip speed ratio, and there is an increase in the turbulence levels just after the location of X_t, where ΔU decreases rapidly.

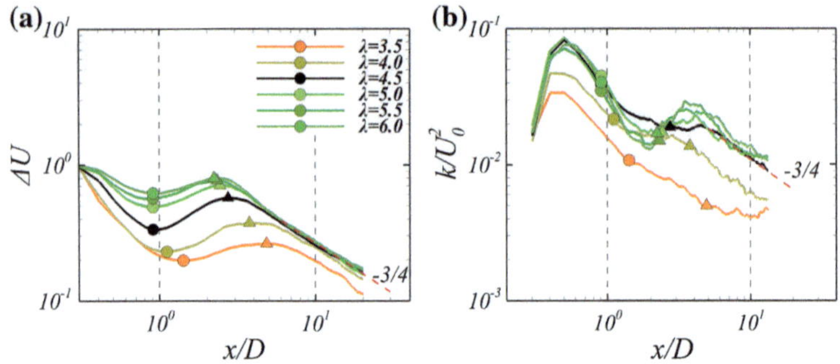

Fig. 4 Centreline profiles of velocity deficit (**a**) and turbulent kinetic energy (**b**) for tip speed ratios $\lambda = [3.5, 6.0]$

Fig. 5 Correlation between velocity deficit and turbulent kinetic energy values for x/D > 5 with the turbine rotating at $\lambda = 4.5$. Symbols coloured as a function of streamwise distance x/D

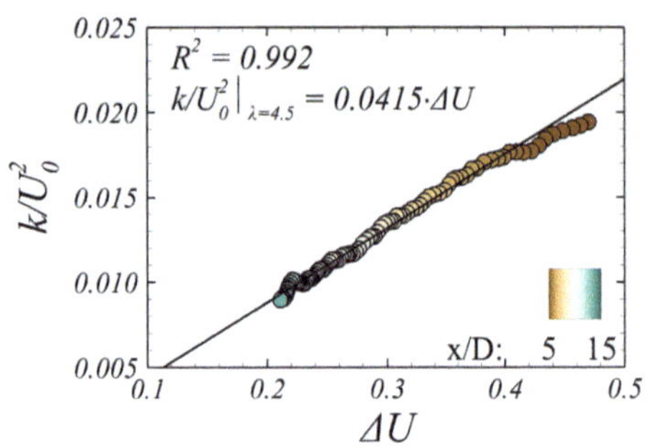

The LES-predicted velocity deficit and turbulent kinetic energy after X_t decay proportionally to the downstream distance with the exponent $n = -3/4$. Additionally, for $\lambda \geq 4.5$ both velocity deficit and turbulent kinetic energy feature similar values indicating that the far-wake dynamics do not considerably change depending on the operational regime of the turbine once this is at or above the optimum tip speed ratio. Such correlation of the rate of wake recovery means that the far-wake can be modelled with a power law such as,

$$\Delta U = A(x/D)^n \tag{6}$$

with $n = -3/4$ and $A = 1.45$, both obtained from fitting of the data for x/D > 5. These values are well within the interval of $A \in [1, 3]$ and $n \in [-5/4, -1/4]$ normally considered for wind turbine wakes [15].

Figure 5 presents the correlation between ΔU and k for the values obtained by LES-ALM after x/D = 5. As expected from Fig. 4, the same decay rate of these variables leads to their quasi-linear correlation which, for the case of $\lambda = 4.5$, is $k/U_0^2 = 0.0415 \cdot \Delta U$, obtaining a coefficient of determination $R^2 = 0.992$.

3.4 Transversal Wake Recovery

The design of arrays comprising multiple rows of turbines is often done using analytical engineering-precision models [3] that are computationally inexpensive, while offering a fairly good accuracy. Stallard et al. [2] analysed the properties of the wake behind the tidal turbine and found it evolves similarly to a planar wake, and thus proposed an analytical model for the velocity deficit at any given transverse profile for x/D > 8. This model considers the maximum velocity deficit ΔU_{max} and $y_{1/2}$ that is the transverse location from the centreline at which ΔU is equal to $\Delta U_{max}/2$. Considering the turbine rotating at its optimum tip speed ratio, the wake parameters calculated by [2] are,

$$\Delta U_{max}/U_0 = 0.864(x/D)^{(p-1)} - 0.126 \quad (7)$$

$$y_{1/2}/R = 0.412(x/D)^p + 0.5 \quad (8)$$

with $p = 1/2$ being the rate of decay. Once ΔU_{max} and $y_{1/2}$ are determined, the model velocity deficit distribution follows a Gaussian distribution as,

$$\Delta U/\Delta U_{max} = e^{-\left(y^2/y_{1/2}^2\right)^{\alpha}} \quad (9)$$

with α equal to $ln(2)$. For the purpose of validation completeness, the profiles of velocity deficit at $\lambda = 4.5$ from the LES-ALM simulation at distances between 5 <x/D< 30 every 5D are compared to the model (Eq. 9) in Fig. 6. The normalised velocity deficit profiles show a good agreement with the wake model until x/D = 20 with an overestimation of the wake width observed for x/D = 25 and 30.

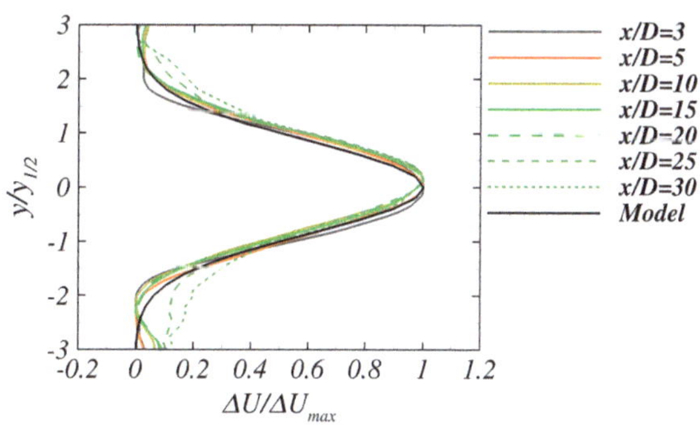

Fig. 6 Transverse distribution of $\Delta U/\Delta U_{max}$ over $y/y_{1/2}$ at locations x/D = 3, 5, 10, 15, 20, 25 and 30 with the turbine rotating at the optimum tip speed ratio compared to the Gaussian profile of Eq. 9

4 Conclusions

An actuator line model has been implemented into an in-house large-eddy simulation solver, and has been employed for the simulation of a tidal stream turbine operating at a range of tip speed ratios. LES-ALM predictions of thrust and power coefficients were compared with experimental data and reasonable agreement was obtained, improving the predictions of previous simulations using a RANS model. LES-ALM-predicted velocity deficit and turbulence intensity profiles in the wake of the turbine were validated against experiments demonstrating that the method is able to replicate the wake downstream of the turbine. The wake is characterised by a near-wake region of low-velocities and high-turbulence. In the analysis of the rate of momentum recovery along the wake for a range of tip speed ratios, it was found that, once the minimum velocity deficit is reached, both velocity deficit and turbulent kinetic energy in the far-wake recovered at a constant exponential rate of $-3/4$. It has been shown that a linear correlation between wake deficit and turbulent kinetic energy takes place in the far-wake region. Furthermore, transverse profiles of the velocity deficit were compared to the profiles of an analytical model and excellent agreement was achieved. Overall, this study has provided evidence that the wake downstream of a horizontal axis tidal turbine can be predicted with good accuracy by the proposed LES-ALM method, indicating its suitability for the future simulations of arrays of turbines.

Acknowledgements The authors would like to acknowledge the support of the Supercomputing Wales project, which is part-funded by the European Regional Development Fund (ERDF) via the Welsh Government.

References

1. Ouro P, Harrold M, Stoesser T, Bromley P (2017) Hydrodynamic loadings on a horizontal axis tidal turbine prototype. J Fluids Struct 71:78–95
2. Stallard T, Feng T, Stansby PK (2015) Experimental study of the mean wake of a tidal stream rotor in a shallow turbulent flow. J Fluids Struct 54:235–246
3. Stansby P, Stallard T (2016) Fast optimisation of tidal stream turbine positions for power generation in small arrays with low blockage based on superposition of self-similar far-wake velocity deficit profiles. Renew Energy 92:366–375
4. Breton S, Sumner J, Sorensen J, Hansen K, Sarmast S, Ivanell S (2017) A survey of modelling methods for high-fidelity wind farm simulations using large eddy simulation. Philos Trans A 375:20160,097
5. Munters W, Meyers J (2018) Dynamic strategies for yaw and induction control of wind farms based on large-eddy simulation and optimization. Energies 11(1):177
6. Ouro P, Stoesser T, Fraga B, Lopez-Novoa U (2018) Hydro3D
7. Cevheri M, McSherry R, Stoesser T (2016) A local mesh refinement approach for large-eddy simulations of turbulent flows. Int J Numer Methods Fluids 82:261–285
8. Ouro P, Stoesser T (2017) An immersed boundary-based large-eddy simulation approach to predict the performance of vertical axis tidal turbines. Comput Fluids 152:74–87
9. Yang X, Howard KB, Guala M, Sotiropoulos F (2015) Effects of a three-dimensional hill on the wake characteristics of a model wind turbine. Phys Fluids 27(2):025,103

10. Shen WZ, Mikkelsen R, Sørensen JN, Bak C (2005) Tip loss corrections for wind turbine computations. Wind Energy 8(4):457–475
11. Martínez-Tossas LA, Churchfield MJ, Yilmaz AE, Sarlak H, Johnson PL, Sørensen JN, Meyers J, Meneveau C (2018) Comparison of four large-eddy simulation research codes and effects of model coefficient and inflow turbulence in actuator-line-based wind turbine modeling. J Renew Sustain Energy 10(3), 033301
12. Olczak A, Stallard T, Feng T, Stansby PK (2016) Comparison of a RANS blade element model for tidal turbine arrays with laboratory scale measurements of wake velocity and rotor thrust. J Renew Sustain Energy 64:87–106
13. Park J, Basu S, Manuel L (2014) Large-eddy simulation of stable boundary layer turbulence and estimation of associated wind turbine loads. Wind Energy 17(3):359–384
14. Kang S, Yang XL, Sotiropoulos F (2014) On the onset of wake meandering for an axial flow turbine in a turbulent open channel flow. J Fluid Mech 744:376–403
15. Iungo GV, Porté-Agel F (2014) Volumetric lidar scanning of wind turbine wakes under convective and neutral atmospheric stability regimes. J Atmos Ocean Technol 31(10):2035–2048

Harmonic Balance Navier–Stokes Analysis of Tidal Stream Turbine Wave Loads

A. Cavazzini, M. S. Campobasso, M. Marconcini, R. Pacciani and A. Arnone

Abstract ARCTIC, a novel incompressible Reynolds–averaged Navier–Stokes finite volume code for the hydrodynamic analysis of open rotor unsteady loads is presented. One of its unique features is a harmonic balance solver enabling high–fidelity analyses of turbine periodic hydrodynamic loads with runtimes reduced by more than one order of magnitude over conventional time–domain CFD, and with negligible accuracy penalty. The strength of the new technology is demonstrated by analyzing with both harmonic balance and time–domain solvers the load fluctuations of a realistic tidal stream turbine. Such fluctuations are caused by a harmonic perturbation of the freestream velocity similar to that due to surface gravity waves.

1 Introduction

Horizontal axis tidal stream turbines (TSTs) can be used to harvest renewable tidal stream energy, but their operation in harsh marine environments poses design challenges due to unsteady loads resulting in fatigue and/or extreme loads [1]. Loads due to current speed fluctuations resulting from surface gravity waves are predominantly

A. Cavazzini · M. S. Campobasso (✉)
Department of Engineering, Lancaster University, Gillow Avenue,
Lancaster LA1 4YW, UK
e-mail: m.s.campobasso@lancaster.ac.uk

A. Cavazzini
e-mail: a.cavazzini@lancaster.ac.uk

M. Marconcini · R. Pacciani · A. Arnone
Department of Industrial Engineering, University of Florence, Via Santa Marta 3,
50139 Florence, Italy
e-mail: michele.marconcini@unifi.it

R. Pacciani
e-mail: roberto.pacciani@unifi.it

A. Arnone
e-mail: andrea.arnone@unifi.it

E. Ferrer and A. Montlaur (eds.), *Recent Advances in CFD for Wind and Tidal Offshore Turbines*, Springer Tracts in Mechanical Engineering,
https://doi.org/10.1007/978-3-030-11887-7_4

periodic in several design–driving conditions, and their reliable prediction, at design stage, can be accomplished using Navier–Stokes Computational Fluid Dynamics (CFD). The time–domain (*TD*) CFD solution of periodic rotor flows can be very time–consuming, whereas the frequency–domain harmonic balance (*HB*) solution of the Reynolds–averaged Navier–Stokes (RANS) equations [2] of such flows can reduce runtimes by more than one order of magnitude with no significant accuracy loss, as demonstrated for the horizontal axis wind turbine case using compressible flow CFD [3]. To the best of the authors' knowledge, this study presents the first industrial–scale application of the *HB* RANS technology to an incompressible periodic flow, namely the analysis of the unsteady hydrodynamic loads of a TST rotor subject to large harmonic variations of the oncoming current speed. The *TD* and *HB* governing equations are provided in Sect. 2; the numerical methods used by the new incompressible finite volume RANS code ARCTIC are in Sect. 3. The TST test case is analyzed in Sect. 4, and a summary is given in Sect. 5.

2 Governing Equations

Turbulent incompressible flows can be determined by solving the RANS equations, featuring the Reynolds stress tensor in the momentum equations, and using Menter's Shear Stress Transport (SST) turbulence model for the turbulence closure. When solving rotor flow problems, it is convenient to formulate the governing equations in a rotating frame of reference, since, in the absence of external hydrodynamic forcing (e.g. yawed and/or sheared current) and instabilities (e.g. rotor blade dynamic stall), this enables solving a problem that is unsteady in the absolute (stationary) frame as a steady problem in the relative (rotating) frame. Using this approach, the grid position is fixed during the analysis. The unknown flow velocity components in ARCTIC are those at an arbitrary time in the absolute frame, even though the conservation laws are expressed in the rotating frame. The governing equations in the time– and frequency–domain are discussed in Sects. 2.1 and 2.2, respectively.

2.1 Time–Domain Equations

The Arbitrary Lagrangian–Eulerian (ALE) integral form of the system of the time–dependent RANS and SST equations is written in a rotating Cartesian coordinate system, rotating about the z-axis with constant angular speed $\underline{\Omega}$, and with the y-axis usually being the blade axis. Given a control volume C with boundary S, the *TD* ALE integral form of the system of the RANS and SST equations is:

$$\frac{\partial}{\partial t}\left(\int_C \mathbf{U}' dC\right) + \oint_S (\underline{\mathbf{\Phi}}_c - \underline{\mathbf{\Phi}}_d) \cdot d\underline{S} - \int_C \mathbf{S}_r dC = 0 \tag{1}$$

with $\mathbf{U}' = [0 \quad \underline{\mathrm{u}}^T \quad k \quad \omega]^T$. The symbols $\underline{\mathrm{u}}$, k, and ω denote, respectively, the flow velocity vector, the turbulent kinetic energy per unit mass, and the specific dissipation rate of k. The array of convective fluxes is:

$$\mathbf{\Phi}_c = \begin{bmatrix} \left(\underline{\mathrm{u}} - \underline{\mathrm{u}}_b\right)^T \\ \left(\underline{\mathrm{u}} - \underline{\mathrm{u}}_b\right)^T \underline{\mathrm{u}} + pI \\ k\left(\underline{\mathrm{u}} - \underline{\mathrm{u}}_b\right)^T \\ \omega\left(\underline{\mathrm{u}} - \underline{\mathrm{u}}_b\right)^T \end{bmatrix} \tag{2}$$

where I is the (3×3) identity matrix, and p is the static pressure. The boundary velocity $\underline{\mathrm{u}}_b$ is given by:

$$\underline{\mathrm{u}}_b = \underline{\Omega} \times \underline{\mathrm{r}} \tag{3}$$

in which $\underline{\mathrm{r}}$ is the position vector. The definition of the generalized diffusive flux vector $\underline{\mathbf{\Phi}}_d$ is reported in [4], and the source term $\mathbf{S}_r$ is given by:

$$\mathbf{S}_r = [0 \quad -\Omega v \quad \Omega u \quad 0 \quad S_k \quad S_\omega]^T \tag{4}$$

where Ω is the modulus of $\underline{\Omega}$. The symbols S_k and S_ω denote respectively the source terms of the k and ω equations of the SST turbulence model, and are defined in [4]. The equations of System (1) are nondimensionalized using the freestream density (ρ_∞), and absolute static pressure (p_∞), and a reference length l_r. Reference speed and time are respectively $v_r = \sqrt{p_\infty/\rho_\infty}$ and $t_r = l_r/v_r$.

2.2 Harmonic Balance Equations

The derivation of the frequency–domain *HB* RANS and SST equations in a rotating frame follows the steps of that in the inertial frame [2, 5], summarized below. The sought periodic flow field due to an external excitation of known angular frequency Ω_e is written as a truncated Fourier series with $(2N_H + 1)$ components, namely N_H sinusoidal components, N_H cosinusoidal components, and a time–independent term, with N_H being a user–given parameter. The Fourier series expansion is then inserted in the *TD* equations (System (1)). After suitable rearrangements [5], this results in the original system of $N_{pde} = 6$ time–dependent partial differential equations (PDEs) becoming a system of $[N_{pde} \times (2N_H + 1)]$ time–independent PDEs, the solution of which yields the $(2N_H + 1)$ components of the truncated Fourier series. Writing the *HB* RANS and turbulence model equations in the Fourier space, however, is challenging, due to the high level of nonlinearity of the equations involved; it was found more convenient to re–cast the *HB* CFD equations in the TD [2]. Re–casting the *HB* equations in the time–domain, results in the *HB* equations becoming a system of $[N_{pde} \times (2N_H + 1)]$ steady flow problems, and the *HB* solution becoming a set of equally spaced flow states or snapshots of the unknown periodic flow. The *HB* form of the RANS and SST equations is found to be:

$$\Omega_e D\left(\int_{C_H} \mathbf{U}'_H \, dC_H\right) + \oint_{S_H} (\underline{\mathbf{\Phi}}_{cH} - \underline{\mathbf{\Phi}}_{dH}) \cdot d\underline{S}_H - \int_{C_H} \mathbf{S}_H dC_H = 0 \quad (5)$$

The array $\mathbf{U}'_H$ is made up of $2N_H + 1$ periodic flow snapshots at $2N_H + 1$ equally spaced times t_n given by:

$$t_n = \frac{n}{(2N_H + 1)} \frac{2\pi}{\Omega_e}, \qquad n = 0, 1, \ldots, 2N_H \quad (6)$$

The structure of the array $\mathbf{U}_H$ is thus: $\mathbf{U}'_H = [\tilde{\mathbf{U}}'(t_0)^T \; \tilde{\mathbf{U}}'(t_1)^T \ldots \tilde{\mathbf{U}}'(t_{2N_H})^T]^T$, and is the same of that of all other variables with a subscript H appearing in Eq. (5). The spectral operator D is a $[(2N_H + 1) \times (2N_H + 1)]$ antisymmetric matrix [3, 5], which couples all $(2N_H + 1)$ flow snapshots.

3 Numerical Method

The ARCTIC finite volume CFD code solves the *TD* and *HB* RANS and SST equations using an explicit multigrid algorithm similar to that of the compressible finite volume COSA code [4], based on Runge–Kutta (RK) pseudo–time–marching. The lack of a physical time–derivative in the continuity equation is dealt with by adopting Chorin's artificial compressibility [6], using which System (1) becomes:

$$\Gamma^{-1}\frac{\partial}{\partial \tau}\left(\int_C \mathbf{U} dC\right) + \frac{\partial}{\partial t}\left(\int_C I^* \mathbf{U} dC\right) + \oint_S (\underline{\mathbf{\Phi}}_c - \underline{\mathbf{\Phi}}_d) \cdot d\underline{S} - \int_C \mathbf{S}_r dC = 0 \quad (7)$$

where $\mathbf{U} = [p \;\; \underline{\mathrm{u}}^T \;\; k \;\; \omega]^T$, $\Gamma^{-1} = diag\left(1/\beta^2, 1, 1, 1\right)$, β is a user–given constant, τ is the pseudo–time, and I^* is the (4×4) identity matrix with the first entry set to 0.

3.1 Space Discretization

ARCTIC solves the RANS and SST equations using structured multi–block grids. The convective fluxes per unit area $\mathbf{\Phi}_{cf}$ along the cell face normal unit vector $\underline{n}$ are computed using van Leer's second order MUSCL extrapolations and Roe's flux difference splitting, i.e.:

$$\mathbf{\Phi}^*_{cf} = \frac{1}{2}\left[\mathbf{\Phi}_{cf}\left(\mathbf{U}_L\right) + \mathbf{\Phi}_{cf}\left(\mathbf{U}_R\right) - \left|\frac{\partial \mathbf{\Phi}_{cf}}{\partial \mathbf{U}}\right| \delta \mathbf{U}\right] \quad (8)$$

where the superscript $*$, and the subscripts f, L and R denote respectively numerical approximation, face value, and values to the left and the right of the face, and $\delta\mathbf{U} = \mathbf{U}_R - \mathbf{U}_L$. The term $\left|\partial\boldsymbol{\Phi}_{cf}/\partial\mathbf{U}\right|\delta\mathbf{U}$ is the numerical dissipation, whose construction requires diagonalizing the Jacobian $\left(\partial\boldsymbol{\Phi}_{cf}/\partial\mathbf{U}\right)$. Since, unlike in the case of the compressible RANS and SST equations [7], the Jacobians of the RANS and SST equations are decoupled in the incompressible case, only the Jacobian of the RANS equations is considered below, namely:

$$\frac{\partial\boldsymbol{\Phi}_{cf}}{\partial\mathbf{U}} = \begin{bmatrix} 0 & \beta^2 n_x & \beta^2 n_y & \beta^2 n_z \\ n_x & u n_x + u_n - u_{bn} & u n_y & u n_z \\ n_y & v n_x & v n_y + u_n - u_{bn} & v n_z \\ n_z & w n_x & w n_y & w n_z + u_n - u_{bn} \end{bmatrix} \tag{9}$$

where $u_n = u n_x + v n_y + w n_z$ and $u_{bn} = u_b n_x + v_b n_y + w_b n_z$ are, respectively, the fluid and the cell face velocity components along $\underline{\mathrm{n}}$.

The numerical dissipation of the RANS equations can be shown to be:

$$\delta\boldsymbol{\Phi} = \sum_{k=1}^{4} |\lambda_k| \delta\mathbf{W}_k R_k \tag{10}$$

where

$$\lambda_{1,2} = u_n - u_{bn} \quad \lambda_3 = u_n - \frac{u_{bn}}{2} + c \quad \lambda_4 = u_n - \frac{u_{bn}}{2} - c \tag{11}$$

are the eigenvalues of $\left(\partial\boldsymbol{\Phi}_{cf}/\partial\mathbf{U}\right)$, $c = \sqrt{\beta^2 + (u_n - u_{bn}/2)^2}$ is the artificial speed of sound, $\delta\mathbf{W}_k = R^{-1}\delta\mathbf{U}$ is the variation of the characteristic variable associated with the eigenvalue λ_k, and R_k and R^{-1} are respectively the kth right eigenvector and the matrix of left eigenvectors of $\left(\partial\boldsymbol{\Phi}_{cf}/\partial\mathbf{U}\right)$. The discretization of the diffusive fluxes and the source terms resulting from the production terms of the SST model is based on second order central differencing [4]. The implementation of the freestream boundary conditions is based on the characteristics of the Jabobian in Eq. (9), and the appropriate characteristics and freestream data to be used at inflow and outflow boundaries are selected automatically depending on whether the flow enters or leaves the domain. The cell face velocities given by Eq. (3) are computed using the freestream–capturing geometric formulation of [8].

3.2 *Numerical Integration*

The *TD* equations are solved using Jameson's dual-time–stepping. Denoting by $\mathbf{R}_{\Phi}$ the array of cell residuals resulting from the balance of the convective and diffusive fluxes over the cell faces, and all source terms over the cell volume, and $\mathbf{Q}$ the discrete counterpart of the array $\mathbf{U}$ evaluated at the cell center at physical time level

$n+1$, and adopting a second order backward difference to discretize the physical time–derivative of $\mathbf{U}$, the semi–discrete counterpart of System (7) reads:

$$\frac{\partial \mathbf{Q}^{n+1}}{\partial \tau} V + I^* \frac{3\mathbf{Q}^{n+1} - 4\mathbf{Q}^n + \mathbf{Q}^{n-1}}{2\Delta t} V + \mathbf{R}_\Phi(\mathbf{Q}^{n+1}) = 0 \quad (12)$$

where $\mathbf{Q}^n$ and $\mathbf{Q}^{n-1}$ are the known solution arrays at time–levels n and $n-1$ respectively, and V is the cell volume. At each new time level $n+1$, $\mathbf{Q}^{n+1}$ is obtained by marching the equations in pseudo–time towards a steady state, discretizing the pseudo–time–derivative with an explicit four–stage RK scheme, and using local time–stepping, implicit residual smoothing (IRS) and multigrid for convergence acceleration. To avoid instabilities when the physical–time–step Δt is notably smaller than the pseudo–time–step $\Delta\tau$, the term $3\mathbf{Q}^{n+1}/2\Delta t$ in Eq. (12) is treated implicitly in the RK scheme. The mth *TD* RK cycle on the finest grid level is thus:

$$\left[I + 1.5\alpha_m \frac{\Delta\tau}{\Delta t} I^*\right] \mathbf{Q}^m = \mathbf{Q}^0 + 1.5\alpha_m \frac{\Delta\tau}{\Delta t} I^* \mathbf{Q}^{m-1} - \alpha_m \Delta\tau V^{-1} L_{IRS} \mathbf{R}_g(\mathbf{Q}^{m-1}) \quad (13)$$

where α_m is the mth RK coefficient, $\mathbf{Q}^m$ is shorthand for $\mathbf{Q}^{n+1,m}$, $\mathbf{Q}^0$ is the solution at the beginning of the RK cycle, L_{IRS} is the IRS operator, and

$$\mathbf{R}_g(\mathbf{Q}^{m-1}) = I^* \frac{3\mathbf{Q}^{n+1,m-1} - 4\mathbf{Q}^n + \mathbf{Q}^{n-1}}{2\Delta t} V + \mathbf{R}_\Phi(\mathbf{Q}^{n+1,m-1}) \quad (14)$$

Steady problems are solved with RK time–stepping enhanced by the aforementioned acceleration strategies, setting $\Delta\tau/\Delta t = 0$ and $\mathbf{R}_g = \mathbf{R}_\Phi$ in Eq. (13). The smoother of Eq. (13) is for the RANS equations; additional terms appear in the smoother of the SST equations, due to the implicit treatment of the destruction terms [4].

The integration of the *HB* equations is similar to that of steady problems. The *HB* counterpart of the semi–discrete *TD* Eq. (12) is:

$$\frac{\partial \mathbf{Q}_H}{\partial \tau} V_H + D I_H^* \mathbf{Q}_H \Omega_e V_H + \mathbf{R}_{\Phi H}(\mathbf{Q}_H) = 0 \quad (15)$$

in which the array $\mathbf{Q}_H$ has $(2N_H + 1)$ flow states, each referring to the physical times defined by Eq. (6), and thus has length $[N_{pde} \times (2N_H + 1)]$. The array $\mathbf{R}_{\Phi H}$ has the same structure of $\mathbf{Q}_H$, and its $(2N_H + 1)$ states are the residuals associated with the convective fluxes, the diffusive fluxes and all source terms at the considered physical times, whereas the entries of V_H are the cell volumes at the $(2N_H + 1)$ times. The mth *HB* RK cycle on the finest grid level reads:

$$\mathbf{Q}_H^{m+1} = \mathbf{Q}_H^0 + \alpha_m \Delta\tau_H V_H^{-1} L_{IRS,H} \mathbf{R}_{gH}(\mathbf{Q}_H^{m-1}). \quad (16)$$

where the entries of $\Delta\tau_H$ are the local time–steps for the $(2N_H + 1)$ flow states, $L_{IRS,H}$ is a block–matrix with $2N_H + 1$ blocks, each providing the IRS operator for one steady subproblem, and $\mathbf{R}_{gH}(\mathbf{Q}_H) = \Omega_e V_H D I_H^* \mathbf{Q}_H + \mathbf{R}_{\Phi H}(\mathbf{Q}_H)$.

4 Results

The test case considered below is a hypothetical two-blade TST with rotor diameter $D = 20\,\text{m}$ and hub at 40 m depth ($p_\infty = 5\,\text{bar}$), operating with current speed $w_\infty = 2.5\,\text{m/s}$. The tip–speed ratio is $\lambda = 5.4$, and the blade geometry is that of the National Renewable Energy Laboratory Phase VI wind turbine used in a previous COSA compressible flow study [3]. The CFD structured multi–block grid used herein is the same one used in that COSA study, and has about 2.1 M cells for a 180° rotor sector. The nondimensionalized steady flow conditions analyzed herein are the same of the wind turbine in [3], with $\lambda = 5.4$ and Reynolds number $Re = 4.93 \times 10^5$, based on wind freestream velocity and $l_r = 1\,\text{m}$. This grid provided grid–independent solutions for the considered steady flow, with nondimensionalized minimum wall distance y^+ of order 1 at all wall boundaries. To use this grid and the given Re and λ for the TST analysis herein, it was necessary to use a modified kinematic viscosity of $\nu = 1.008 \times 10^{-5}\,\text{m}^2/\text{s}$ and $l_r = 1.988\,\text{m}$. The dimensions of the physical domain, the boundary conditions (BCs) applied to its boundaries, and a view of the hydrofoil mesh are reported in Fig. 1. The TST steady flow and unsteady regime due to a harmonic fluctuation of the current speed, similar to that due to surface gravity waves, are examined in Sects. 4.1 and 4.2, respectively. The 180° rotor sector was used in all cases enforcing periodicity BCs on the lateral boundaries.

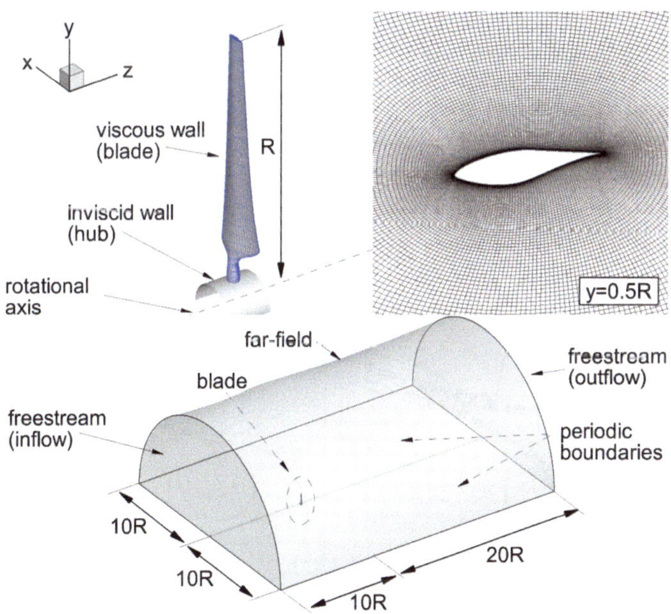

Fig. 1 TST blade geometry (top left), hydrofoil grid at 50% tip radius (top right), and domain dimensions and boundary conditions (bottom)

4.1 Steady Flow Analyses

Here the simulations of the considered TST steady regime based on the new incompressible code ARCTIC and the well validated compressible code COSA [3] are cross–compared. To avoid compressibility effects, a freestream Mach number of 0.0206 and low–speed preconditioning (LSP) [7] are used in the COSA analysis. Denoting respectively by F and T rotor thrust and torque, the thrust coefficient C_F and the torque coefficient C_T are:

$$C_F = \frac{8F}{\rho_\infty w_\infty^2 \pi D^2} \qquad C_T = \frac{16T}{\rho_\infty w_\infty^2 \pi D^3} \tag{17}$$

The pressure and viscous components of C_F and C_T computed by ARCTIC and COSA are compared in Table 1. An excellent agreement is noted, as the maximum difference between corresponding coefficients is lower than 2%.

The ARCTIC and COSA contours of velocity magnitude are compared in Fig. 2, where the left and right plots refer respectively to the blade sections at 30 and 80% rotor radius, and the velocity magnitude is normalized by the freestream value. An excellent agreement between the compressible and incompressible code solutions is observed also at this level, confirming the correctness of the implementation of the steady solver of the new incompressible code.

Table 1 Pressure and viscous components of rotor thrust coefficient C_F and torque coefficient C_T computed by COSA compressible code and ARCTIC incompressible code

	C_{F_p}	C_{F_v}	C_{T_p}	C_{T_v}
COSA	7.54×10^{-1}	3.96×10^{-3}	9.66×10^{-2}	-1.52×10^{-2}
ARCTIC	7.45×10^{-1}	3.88×10^{-3}	9.48×10^{-2}	-1.52×10^{-2}

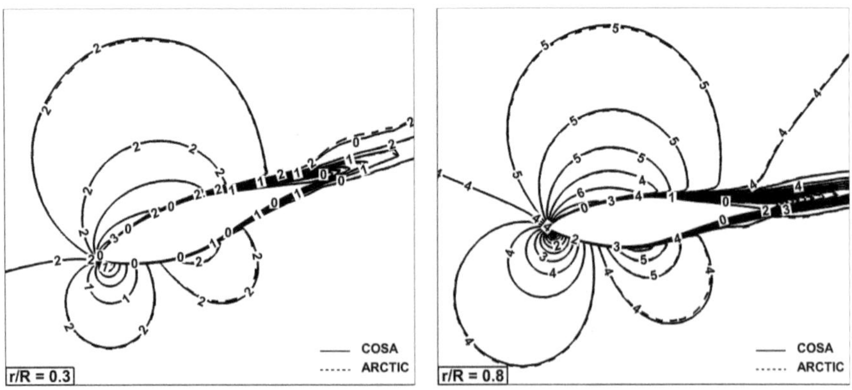

Fig. 2 Contours of velocity magnitude at 30% (left) and 80% (right) rotor radius computed by COSA and ARCTIC steady flow analyses

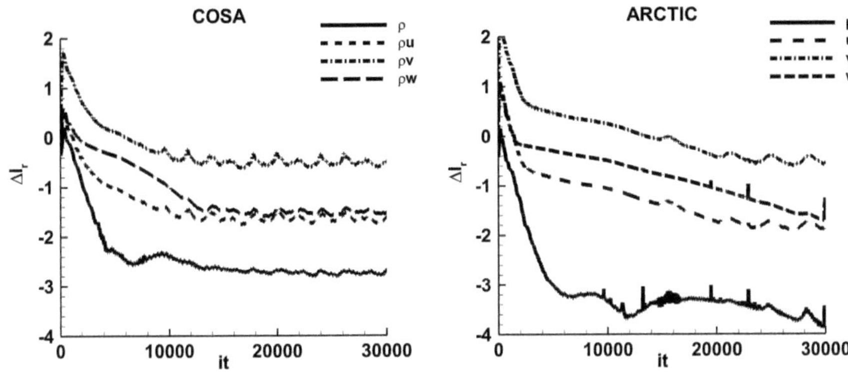

Fig. 3 Continuity and momentum residual convergence histories of steady COSA (left) and ARCTIC (right) solvers

Figure 3 reports the convergence of the residuals of the continuity and momentum equations of COSA (left) and ARCTIC (right). The variable Δl_r on the vertical axes is the logarithm in base 10 of the current residual normalized by the residual at the first iteration. No MG was used for these analyses, and the variable it on the x-axis is the number of single–grid iterations. Overall, the residual convergence of corresponding equations is comparable for the two codes, although the residual of the continuity equation of ARCTIC decreases by more than one order of magnitude over that of COSA, whereas the final residuals of the momentum equations of COSA are marginally lower than those of ARCTIC.

4.2 Unsteady Flow Due to Harmonic Perturbation of Current Speed

Here the unsteady periodic response of TST rotor loads to a harmonic planar perturbation of the freestream current speed is considered. The harmonic perturbation is prescribed as a harmonic variation of the grid velocity along the turbine axis using constant freestream conditions. This is to avoid the high grid refinement upstream of the rotor that would be required to prevent numerical dissipation from damping the propagation of freestream perturbations. The longitudinal grid motion is:

$$w_g(t) = \Delta w_g \cos(\Omega_e t) \tag{18}$$

In the analyses below, $\Delta w_g = 0.4 w_\infty$, $\Omega_e = \Omega$, and the mean flow has the same λ and Re of the steady regime examined in the previous section. The left plot of Fig. 4 depicts the profile of the rotor thrust coefficient C_F over one oscillation period, obtained with *TD* COSA and ARCTIC simulations, and *HB* ARCTIC analyses using $N_H = 1, 3$; the right plot reports instead the periodic profiles of the rotor

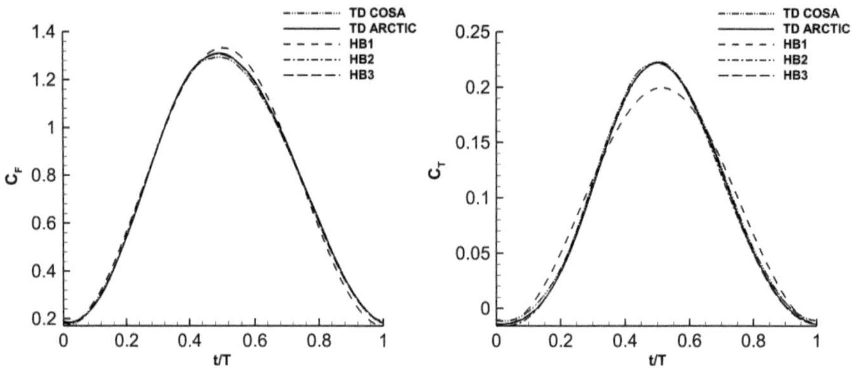

Fig. 4 Periodic profiles of thrust coefficient C_F and torque coefficient C_T obtained with *TD* COSA and ARCTIC simulations, and *HB* ARCTIC simulations

torque coefficient C_T obtained with the same simulations. The peak C_T higher than the theoretical maximum of 1 is merely due to using the mean rather than the instantaneous peak kinetic energy of the current. The *TD* analyses used 360 time–steps per period and 500 RK cycles (no MG was used) per time–step. Thrust and torque were used to monitor the periodicity error. The *TD* solutions were taken to be periodic once, for both outputs, the maximum difference between the instantaneous values of the last two periods was less than 0.5% of the maximum value of the last period, and this happened after 6 periods. The peaks of both C_F and C_T in Fig. 4 are achieved at mid–period because the relative speed of the oncoming flow, resulting by subtracting the negative value of w_g at mid–period (see Eq. (18)) from w_∞, is maximum at this time. This yields close–to–maximum angle of attack and relative speeds along the blade, resulting in the aforementioned peaks. The curves of Fig. 4 show a very good agreement between the *TD* COSA and ARCTIC solutions, which validates the *TD* ARCTIC solver at the integral output level. Small differences are visible only around the peak region of the C_F profiles, possibly due to lower residual convergence of COSA in this region. One also notes that (a) the *HB* ARCTIC profiles of C_F and C_T for $N_H = 2$ (profiles labeled *HB* 2) differ negligibly from the *TD* ARCTIC profiles, and (b) the use of more harmonics does not bring further solution improvement, as the N_H 2 and N_H 3 profiles are superimposed. The harmonic fluctuation of 40% of the current speed results in periodic fluctuations of about 70 and 175% of thrust and torque respectively, indicating power peaks more than twice as high as the steady power. Turbine control or safety system, however, would reduce these fluctuations or shut down the turbine.

The flow detail resolution of the *TD* low–speed compressible and incompressible solvers, and the *HB* 2 incompressible analysis is examined in Fig. 5, comparing the three predictions of the contours of normalized velocity magnitude of the blade sections at 30 and 80% of rotor radius at 60% of the period. The *TD* COSA and ARCTIC solutions differ very little, and no visible difference between the *TD* and *HB* 2 ARCTIC solutions is noted.

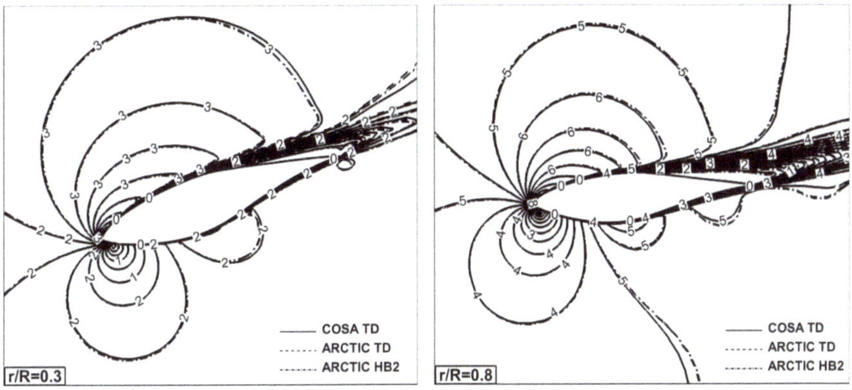

Fig. 5 Contours of velocity magnitude at 30% (left) and 80% (right) rotor radius at 60% of oscillation period computed by *TD* COSA, *TD* ARCTIC, and *HB* 2 ARCTIC analyses

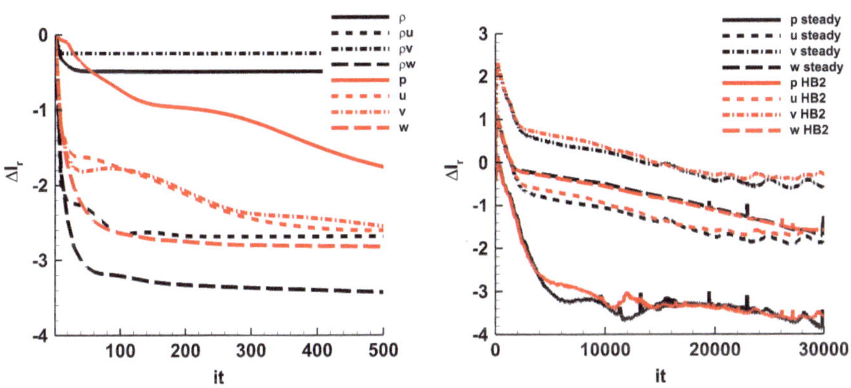

Fig. 6 Continuity and momentum residual convergence histories of *TD* COSA (black) and ARCTIC (red) at 60% of oscillation period (left), and steady and *HB* 2 ARCTIC analyses (right)

The residual convergence histories of the continuity and momentum equations of the *TD* COSA and ARCTIC solvers at 60% of the period are reported in the left plot of Fig. 6 with black and red lines respectively. The residuals of the continuity and the y-momentum equations of the compressible solver stagnate after the first few iterations. The underlying lack of complete convergence may account for the small differences between the COSA and ARCTIC *TD* solutions observed above. The residual converge histories of the steady and *HB* 2 ARCTIC analyses are very similar, as seen in the right plot of Fig. 6, possibly due to the level of unsteadiness being insufficiently high to yield significant differences between the spectral radii of the *HB* and steady iterative solvers [5].

Table 2 provides the runtimes of the *TD* COSA, *TD* ARCTIC AND *HB* ARCTIC analyses in terms of work–units, where one work–unit is the CPU–time needed for 4, 000 RK cycles of the steady ARCTIC solver. One sees that (a) the ARCTIC *HB* 2

Table 2 Computational cost in work–units of *TD* COSA, *TD* ARCTIC and *HB* ARCTIC analyses

TD COSA	*TD* ARCTIC	*HB* 1	*HB* 2	*HB* 3
752	405	25	43	60

solution, which differs negligibly from its *TD* counterpart, is about 9.5x faster than the *TD* analysis, and (b) the incompressible *TD* analysis is 85% faster than the compressible *TD* analysis, as COSA solves one more equation (that for energy) and performs additional operations for LSP.

5 Conclusions

ARCTIC, an incompressible RANS code using Chorin's artificial compressibility was presented. The code has a conventional *TD* solver and a novel frequency–domain *HB* solver. Code validation relied on simulating with ARCTIC and COSA, a well validated LSP–enhanced compressible code, a steady flow condition of a two–blade TST and a periodic flow regime due to a planar harmonic perturbation of the oncoming current. Excellent agreement of blade loads and flow detail was found. Use of the incompressible *HB* solver rather than its *TD* counterpart enabled the CFD analysis of TST periodic hydrodynamic loads to be accelerated by more than 9 times. The approach can be generalized to include additional sources of periodic excitation, including sheared currents and depth–dependent magnitude of oncoming current perturbations due to surface gravity waves.

Acknowledgements This work was supported by the UK Engineering and Physical Sciences Research Council within Lancaster University Impact Acceleration Account, Grant No. EP/R511560/1. The authors are grateful to SIMEC ATLANTIS ENERGY Ltd. for their advice. All ARCTIC simulations were run on the HEC cluster of Lancaster University.

References

1. Amato G, Doyle S, Petley S, Campobasso MS, Milne IA, Aggidis GA (2017) Navier–Stokes CFD analysis of a tidal turbine rotor in steady and planar oscillatory flow. In: European wave and tidal energy conference, Cork, Ireland, September 2017
2. Hall KC, Thomas JP, Clark WS (2002) Computations of unsteady nonlinear flows in cascades using a harmonic balance technique. AIAA J 40(5):879–886
3. Drofelnik J, Da Ronch A, Campobasso MS (2018) Harmonic balance Navier–Stokes aerodynamic analysis of horizontal axis wind turbines in yawed wind. Wind Energy 21(7):515–530
4. Campobasso MS, Piskopakis A, Drofelnik J, Jackson A (2013) Turbulent Navier–Stokes analysis of an oscillating wing in a power-extraction regime using the shear stress transport turbulence model. Comput Fluids 88:136–155
5. Campobasso MS, Baba-Ahmadi MH (2012) Analysis of unsteady flows past horizontal axis wind turbine airfoils based on harmonic balance compressible Navier–Stokes equations with low-speed preconditioning. J Turbomach 134(6)

6. Chorin AJ (1997) A numerical method for solving incompressible viscous flow problems (reprint). J Comput Phys 135:118–125
7. Campobasso MS, Yan M, Bonfiglioli A, Gigante FA, Ferrari L, Balduzzi F, Bianchini A (2018) Low-speed preconditioning for strongly coupled integration of Reynolds-averaged Navier–Stokes equations and two-equation turbulence models. Aerosp Sci Technol 77:286–298
8. Obayashi S (1992) Freestream capturing for moving coordinates in three dimensions. AIAA J 30(4):1125–1128

Analysis of the Aerodynamic Loads on a Wind Turbine in Off-Design Conditions

G. Santo, M. Peeters, W. Van Paepegem and J. Degroote

Abstract In this work, the aerodynamic loads acting on a large horizontal axis wind turbine are analysed in off-design conditions by means of computational fluid dynamics (CFD) simulations. The turbulent wind flow is solved using an unsteady RANS approach and choosing the $k - \epsilon$ model. Appropriate boundary conditions are used in combination with modified wall functions in order to preserve the atmoshperic boundary layer (ABL) profiles throughout the entire domain. An overset technique is used to handle the rotation of the blades throughout the simulated time. Changing both the pitch angle of the blades and the tip-speed ratio (TSR) of the turbine, several operating points are investigated. The performance and the loads are highly affected by the ABL, whose effect is highlighted. The performance of the wind turbine in each simulated operating point is compared to the nominal operating point (NOP). The aerodynamic loads are monitored, analysed and mutually compared throughout the motion of the rotor, in order to identify the most critical conditions for the blade structures.

G. Santo (✉) · J. Degroote
Department of Flow, Heat and Combustion Mechanics, Ghent University,
Sint-Pietersnieuwstraat 41, 9000 Ghent, Belgium
e-mail: gilberto.santo@ugent.be

J. Degroote
e-mail: joris.degroote@ugent.be

M. Peeters · W. Van Paepegem
Department of Materials Science and Engineering, Ghent University,
Technologiepark-Zwijnaarde 903, 9052 Zwijnaarde, Belgium
e-mail: mathijs.peeters@ugent.be

W. Van Paepegem
e-mail: wim.vanpaepegem@ugent.be

E. Ferrer and A. Montlaur (eds.), *Recent Advances in CFD for Wind and Tidal Offshore Turbines*, Springer Tracts in Mechanical Engineering,
https://doi.org/10.1007/978-3-030-11887-7_5

1 Introduction

In the last decades, large efforts have been made to explore alternative techniques to replace fossil fuels as energy source. In particular, among renewable energies, wind energy plays an increasingly important role. Given the aleatory nature of the wind, wind turbines are designed to function in a wide range of operating conditions [1]. For this reason, their operation has to be adapted to face the incoming wind and adapt the output power accordingly. Normally, in large horizontal axis wind turbines, this is achieved by pitching the blades (i.e. rotating them around their own axes) and/or changing the tip speed ratio. These controlling techniques can be investigated in detail by means of computational fluid dynamics (CFD). Sudhamshu et al. [2] carried out CFD (steady RANS) simulations of the NREL Phase VI wind turbine, modelling only one blade and changing both the incoming constant wind speed and the blade pitch angle. Li et al. [3] adopted an overset technique to simulate the aerodynamics of a wind turbine and investigate the effect of various wind speeds and pitch angles on the transient response of the machine. In literature, the effect of the atmospheric boundary layer (ABL) i.e. wind speed increasing with height, is often neglected. Nevertheless, this is expected to induce large differences in the forces acting on each blade during its rotation, also due to the tendency of upscaling the size of modern machines [4].

In this work, a commercial 3-bladed rotor with a diameter of 100 m is modelled by an overset technique, together with its supporting structures, namely tower and nacelle. The whole machine is immersed in the ABL flow, which leads to a wind speed increasing with height. Various operating points in the surrounding of the nominal operating point (NOP) are simulated by changing the pitch angle of the blades and the tip-speed ratio. Both the energy conversion performance of the turbine and the loads acting on the blades will be analysed in detail during the transient rotation of the machine.

2 Methodology

The domain of the flow simulation is displayed in Fig. 1, with indication of the boundary conditions. A distance equal to 5 rotor diameters from the top and side symmetry surfaces is chosen in order to avoid artificial acceleration of the flow. Furthermore, the inflow and the outflow are respectively 5 and 15 rotor diameters away from the rotor.

A 3D fully hexahedral mesh is created for every component of the machine, namely the 3 blades, the hub, the nacelle and the tower. These meshes are then overlapped to a fully structured background mesh and the mutual connectivity is established by means of an overset technique [3, 5]. Figure 2 shows the details of the mesh around each blade, together with an overview of the adopted overset technique. The total number of cells is approximately 56 millions, with 49.6 million cells in

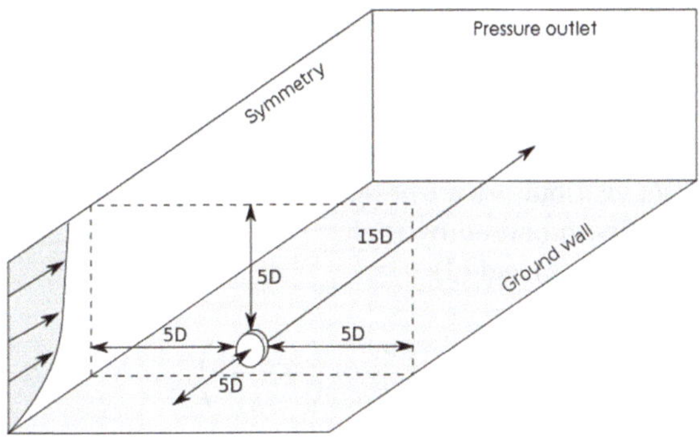

Fig. 1 Simulation layout with the cylinder in the center denoting the domain around the rotor, with diameter D = 100 m

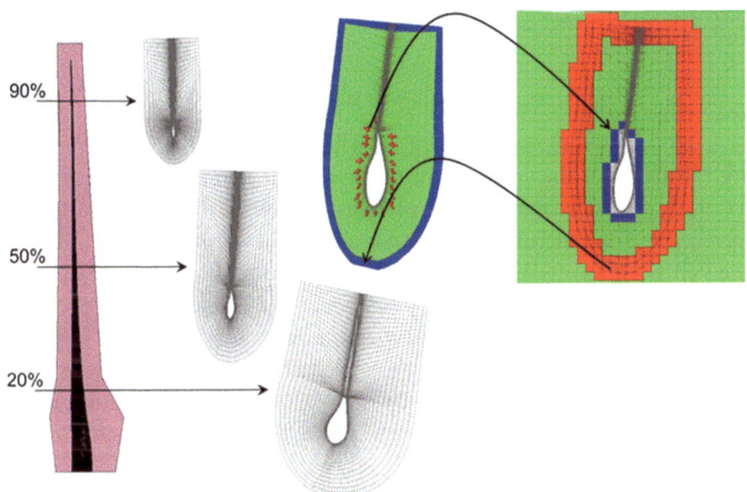

Fig. 2 (left) Sections of the blade component mesh at different spanwise locations and (right) illustration of the mesh connectivity strategy: solve cells are marked in green, donor cells in red and receptor cells in blue

the background mesh and the rest belonging to the component meshes (blades, hub, tower and nacelle).

As an example of how the mesh connectivity is built, the connection of the blade mesh with the background grid is shown in Fig. 2. The background cells encompassed or crossed by the blade walls are deactivated. On the external boundary of the component mesh, the solution is obtained by interpolation from the background mesh. Here, the two meshes are designed to have roughly the same cell size (i.e. edge size around 0.275 m). The (background) cells from where the solution is taken are marked as "donor cells", while the (component) cells receiving solution by interpolation are marked as "receptor cells". At least 4 donor cells contribute to interpolation on each receptor cell.

The fluid is modelled as incompressible and turbulence is represented by means of the $k - \epsilon$ model. At the inlet of the domain, the velocity, turbulent kinetic energy and dissipation rate distributions are prescribed to mimic the distribution of a neutral ABL. The profiles first proposed by Richard and Hoxey [6] are adopted. They are summarized in Eqs. (1), (2) and (3).

$$u(z) = \frac{u_*}{K} ln\left(\frac{z + z_0}{z_0}\right) \quad (1)$$

$$k = \frac{u_*^2}{\sqrt{C_\mu}} \quad (2)$$

$$\epsilon(z) = \frac{u_*^3}{K(z + z_0)} \quad (3)$$

In these equations, u is the wind velocity as a function of the height z, u_* is the friction velocity (an index of the intensity of the wind), z_0 is the aerodynamic roughness length (which provides an estimation of the roughness of the ground wall) and k and ϵ are respectively the turbulent kinetic energy and its dissipation rate. Furthermore, K is the von Karman constant (0.42) and C_μ is a constant of the turbulence model, set to 0.09 [7]. In order to consistently sustain and preserve the inlet profiles across the whole computational domain, a new formulation for the ground wall functions is necessary [8–10]. Thus, following the approach of Parente and Benocci [8], the aerodynamic roughness length is directly included in the wall functions, leading to a modified non-dimensional wall distance z^+ and a modified wall function constant E.

$$z^+_{mod} = \frac{(z + z_0)u_*\rho}{\mu} \quad (4)$$

$$E_{mod} = \frac{\mu}{\rho z_0 u_*} \quad (5)$$

These modified wall functions are used on the ground wall (Fig. 1), while the standard ones are used on every other wall. In this work, the friction velocity is set to $u_* = 0.67$ m/s and the aerodynamic roughness is set to $z_0 = 0.5$ m, leading to a wind speed of 8.5 m/s at the hub height (100 m). This is the nominal wind velocity provided by the blade manufacturer. The turbulent kinetic energy is set to 0.015 m^2/s^2, leading to a turbulence intensity of approximately 1.25% at the hub height. The momentum equations and continuity equation are solved together in a pressure-based solver. Second-order upwind discretization for momentum is applied and a first-order implicit scheme is used for time discretization. The tip-speed ratio (TSR) to be used in this work is defined as the ratio of the speed of the blade tip

Table 1 Analyzed tip speed ratios and respective rotational speed

TSR	Rotational speed
7.5	1.275 rad/s
8.5	1.445 rad/s
9.5	1.615 rad/s

u_{tip} and the undisturbed wind velocity at hub height v_{hub}. The blade speed at the tip can be computed as the product of the rotational speed ω and the blade length R, according to:

$$TSR = \frac{u_{tip}}{v_{hub}} = \frac{\omega R}{v_{hub}} \tag{6}$$

The nominal operating point, as provided by the blade manufacturer, corresponds to a TSR of 8.5 and a pitch angle taken as reference and marked as "pitch 0". Around this point, 8 more points are analyzed, changing both the TSR and the pitch angle of the blades. The rotational speed of the turbine is changed in order to change its tip speed ratios, as summarized by Table 1.

Independently of the TSR to be simulated, each full rotation of the wind turbine rotor is divided into 240 time steps. For each of these TSRs, 3 different blade pitch angles are simulated: the pitch "0°", the pitch "+2°" and the pitch "−2°". The last two are obtained by rotating the blades of +2° or −2° around their axis (positive rotation corresponding to a nose-down rotation of each section). This leads to a total of 9 simulations carried out. First, the turbine rotation is started in the unperturbed ABL, considering a TSR of 8.5 for each analyzed pitch angle. Then, after 5 complete rotations, the TSR is changed to the desired value and additional full rotations (from 2 to 7) are carried out until the torque provided by the machine stabilizes (i.e. difference between the last two thirds of revolution smaller than 1.8%). Finally, only the last revolution is analyzed. Running on 280 cores (10 nodes, each with 2 CPUs of the type 14-core Xeon E5-2680v4, 2.4 GHz, inter-connected via InfiniBand), approximately one day is necessary to perform a complete revolution.

3 Results

In this section, the energy conversion performance of the turbine in each of the simulated operating points will be analyzed, before examining the loads acting on each blade. The torque coefficient c_T and the power coefficient c_P to be used, are defined according to:

$$c_T = \frac{Torque}{\frac{1}{2}\rho v^2 A R} \tag{7}$$

$$c_P = \frac{Power}{\frac{1}{2}\rho v^3 A} = c_T \cdot TSR \tag{8}$$

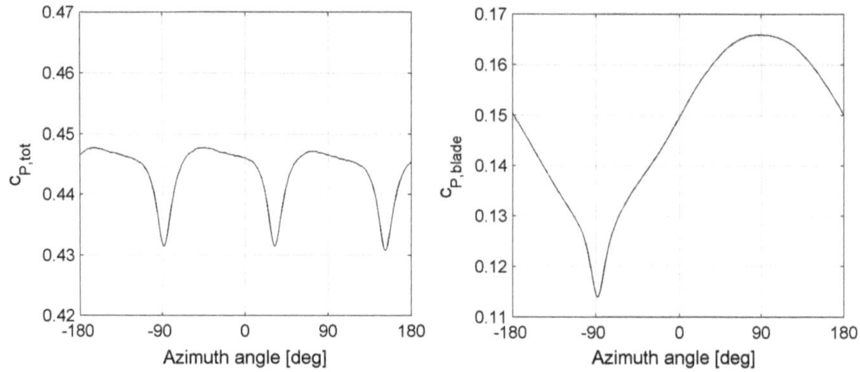

Fig. 3 (left) Total power coefficient and (right) single blade contribution to the total power as a function of the azimuth angle

In these equations, ρ is the constant density of air ($1.225\,\mathrm{kg/m^3}$), v is the undisturbed wind speed at the hub height (8.5 m/s), R the radius of the rotor (50 m) and A its frontal area. Furthermore, to define the position of each blade during the rotation, the 0 azimuth angle position corresponds to the blade horizontally positioned and in uprising motion. Thus, $+90°$ azimuth angle corresponds to the blade pointing upwards and $-90°$ to the blade pointing downwards and passing in front of the tower.

For every operating point, qualitatively the same curve is observed for the extracted power during the last monitored revolution. Figure 3-left reports the total power at the NOP, as a function of the azimuth angle of one of the three blades. On the other hand, Fig. 3-right shows the power contribution of each blade. Note that, in both plots, a drop is visible in the power produced whenever a blade passes in front of the tower. This phenomenon is addressed as "tower dam" and results from the pressure increase induced by the tower obstruction on the suction side of the blade. The single blade contribution is largely influenced by the ABL, producing more power when the blade points upwards (and where the ABL induces higher wind speed) and less when it points downwards (where the ABL induces lower wind speeds). This results in a peak-to-peak amplitude of about 36% of the average value at NOP. Nevertheless, the combination of the three blades produces a stable total power coefficient, which exhibits a maximum deviation from its average lower than 3%.

Qualitatively the same curves can be obtained for every operated point. Figure 4-left shows the average power coefficient obtained for all the examined operating points. It is reported that the operating point corresponding to $+2°$ pitch and 9.5 TSR has a higher power coefficient (+4.7%) than the nominal operating point provided by the manufacturer. When simulating these two particular operating points, the difference in power coefficient between the last two third of revolution (used to assess convergence towards time regime) is, in both cases, lower than 0.3%. It is therefore concluded that the differences observed are larger than the margin of error of the simulations. On the other hand, every other operating point exhibits a lower coefficient (up to -18.6% for $-2°$ pitch and 9.5 TSR) with respect to the NOP.

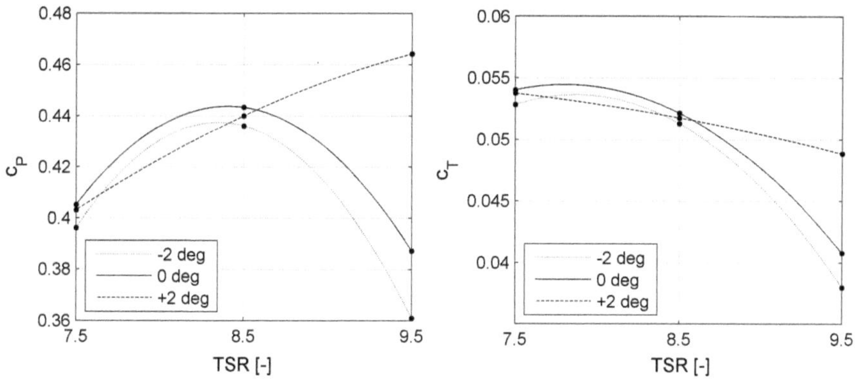

Fig. 4 Average values of (left) total power coefficient and (right) total torque coefficient of the turbine for all the simulated operating points

However, Fig. 4-right reports the torque coefficient corresponding to all the analyzed operating points. From Eq. 8 it can be seen that the power coefficient results from the product of the TSR and the torque coefficient. For every pitch angle, an increase in TSR leads to a decrease in torque coefficient, since the angle of attack on the blade decreases. When the pitch angle is changed, the distribution of the angle of attacks over the entire blade span is shifted. This leads to a different operation of each airfoil lofted along the blade and, additionally, to different directions of the produced lift and drag forces. Consequently, the (positive) contribution of the lift and the (negative) contribution of the drag will combine differently to the produced torque, according to the adopted pitch angle and TSR. The best performing point visible in Fig. 4-left provides a lower torque coefficient compared to the NOP. Nevertheless, the increase in rotational speed (i.e. in TSR) is stronger than the reduction in c_T, leading to an overall higher power coefficient.

A similar analysis can be carried out for the axial load acting on each blade. The axial force is in fact the highest force component and it is the one dominantly contributing to the deflection of the blades during the operation of the machine. The axial force acting on each blade can be related directly to its position. Being largely influenced by the ABL and by the tower-dam effect, a similar plot of Fig. 3-right can be obtained also for the axial force acting on each blade. Figure 5-left shows the average value of the axial force, while Fig. 5-right shows the peak-to-peak amplitude, which directly relates to the arising of fatigue problems. Note that, differently than what is observed for the torque, the average axial force consistently increases with increasing TSR, following the increase in incoming velocity magnitude more than the decrease in angle of attack. A similar behavior is observed also by other authors [11]. The point previously highlighted for having a higher c_P than the NOP also shows a slightly smaller average axial force (-1.7%), but, at the same time, is also the one with the highest amplitude in its oscillation. The operating point with $-2°$ pitch and 9.5 TSR has the highest average value of the axial force (+24.7% when compared to NOP).

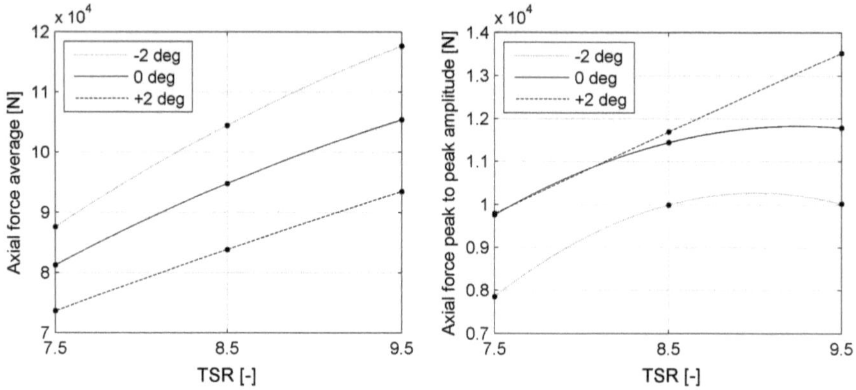

Fig. 5 (left) Average values and (right) peak to peak amplitudes of the axial force acting on each blade

Fig. 6 Distributions of axial force per meter of blade as a function of the blade span, at 90° azimuth angle, namely the highest load condition

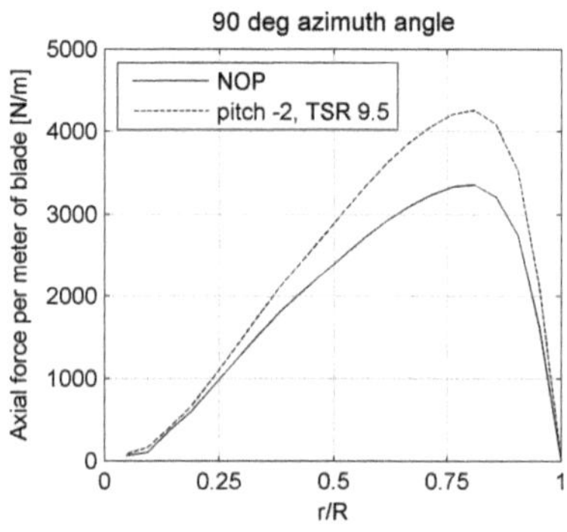

The distribution of the axial force throughout the blade can also be analyzed as done in Fig. 6. A consistent increase of the axial force solicitation is monitored throughout the entire blade span and the highest increase is observed at about 75% of the radius, where the axial force also reaches its peak. A similar condition is reported during the entire rotation.

4 Conclusions

Changing the blade pitch angle and its tip speed ratio, 9 operating points were simulated, keeping the ABL wind flow unchanged. In this way, the characteristic of the turbine has been built in the proximity of the nominal operating point (NOP). A better performing point could be identified.

The axial force average value and amplitude of oscillation were also monitored, showing that the previously identified better performing point is characterized by a higher amplitude of the oscillating axial force on each blade. Furthermore, one of the operating points exhibited an average value of the axial force 24.7% higher than at the NOP. This increase is spread over the entire blade span, reaching its maximum at about 75% of its span.

Following the present work, the operating point with the highest observed axial force will be simulated in a fluid-structure interaction framework developed by the same authors [12], in order to investigate the structural response of the blades in unfavorable off design conditions.

References

1. Hau E (2006) Wind turbines: fundamentals, technologies, application, economics, 2nd edn. Springer, Berlin
2. Sudhamsu AR, Pandey MC et al (2016) Numerical study of effect of pitch angle on performance characteristics of a HAWT. Eng Sci Technol Int J 632–641
3. Li Y, Paik KJ, Xing T, Carrica PM (2012) Dynamic overset CFD simulations of wind turbine aerodynamics. Renew Energy 37(1):285–298
4. Caduff M, Huijbregts MAJ, Althaus HJ, Koehler A, Hellweg S (2012) Wind power electricity: the bigger the turbine, the greener the electricity? Environ Sci Technol 46(9):4725–4733
5. Zahle F, Sørensen NN, Johansen J (2009) Wind turbine rotor-tower interaction using an incompressible overset grid method. Wind Energy 12(6):594–619
6. Richards PJ, Hoxey RP (1993) Appropriate boundary conditions for computational wind engineering models using the $k-\epsilon$ turbulence model. J Wind Eng Ind Aerodyn 46–47:145–153
7. Launder BE, Spalding DB (1972) Lectures in mathematical models of turbulence. Academic Press, London
8. Parente A, Gorlè C, van Beeck J, Benocci C (2011) A comprehensive modelling approach for the neutral atmospheric boundary layer: consistent inflow conditions, wall function and turbulence model. Bound Layer Meteorol 140:411
9. Blocken B, Stathopoulos T, Carmeliet J (2017) CFD simulation of the atmospheric boundary layer: wall function problems. Atmos Environ 41(2):238–252
10. Parente A, Gorlè C, van Beeck J, Benocci C (2011) Improved $k-\epsilon$ model and wall function formulation for the RANS simulation of ABL flows. J Wind Eng Ind Aerodyn 99(4):267–278
11. Dai L, Zhou Q, Zhang Y, Yao S, Kang S, Wang X (2017) Analysis of wind turbine blades aeroelastic performance under yaw conditions. J Wind Eng Ind Aerodyn 171:273–287
12. Santo G, Peeters M, Van Paepegem W, Degroote J (2017) Transient aeroelastic simulations of wind turbines with composite blades. In: proceedings of the 7th GACM colloquium on computational mechanics for young scientists from academia and industry

An Algorithm for the Generation of Biofouled Surfaces for Applications in Marine Hydrodynamics

Sotirios Sarakinos and Angela Busse

Abstract The adverse effects of marine biofouling on marine renewable energy devices are well established. In recent fundamental investigations on fluid flow over this type of surface roughness, marine biofouling has mainly been realized as ordered arrangements of roughness elements. These surfaces cannot be compared to realistic biofouled surfaces which show an irregular distribution of roughness features. In this work, a geometric algorithm for generating realistic surface roughness due to barnacle settlement is presented. The algorithm mimics the settlement behaviour of barnacles and allows the generation of a range of fouling states from very sparse rough surfaces to surfaces that are fully covered by barnacle colonies. The generated surfaces can be used in various applications, e.g. in CFD simulations to establish the fluid dynamic roughness effect of different fouling states or as 3D printed surface tiles for use in wind-tunnel and towing tank experiments.

1 Introduction

The accumulation of marine organisms on ship hulls or underwater surfaces of man-made structures is collectively defined as marine biofouling. The negative effects these colonies of marine organisms cause can be of economic, environmental, and engineering nature. In 2011 Schultz et al. [1] calculated the economic impact of marine biofouling on naval surface ships and noted that the yearly maintenance and operational costs for a ship can increase by millions of dollars with a small change in their biofouling rating. The increased cost in most cases is a result of the excessive fuel needs, caused by a significant increase in the frictional resistance of submerged surfaces due to biofouling [2]. Loxton et al. [3] reported the effect of marine biofouling on marine renewable energy devices. They stated that biofouling

S. Sarakinos (✉) · A. Busse
University of Glasgow, Glasgow G12 8QQ, UK
e-mail: sotirios.sarakinos@glasgow.ac.uk

A. Busse
e-mail: angela.busse@glasgow.ac.uk

E. Ferrer and A. Montlaur (eds.), *Recent Advances in CFD for Wind and Tidal Offshore Turbines*, Springer Tracts in Mechanical Engineering,
https://doi.org/10.1007/978-3-030-11887-7_6

can reduce the energy extraction efficiency of wave or tidal energy harvesting devices, as it can increase the effective diameter of their components, causing an increase in drag and inertia. The need for further study of the effects of biofouling on sea-faring vessels and marine renewable energy devices is apparent.

Marine biofouling can be roughly categorized into microscopic, also known as microfouling, and macroscopic biofouling, also known as macrofouling. The former category encapsulates organisms that are not visible to the naked eye, such as bacteria, diatoms, and protozoa, as well as inorganic matter, such as mud, sand and other materials in particle form [4]. Bacterial organisms and diatoms are responsible for producing a thin film of slime on submerged surfaces. Together with protozoa, rotifers and inorganic material, they form what is known as biofilm. With the term macrofouling all organisms or clusters of organisms are categorized that are large enough to be identified without optical aid. Among the macrofouling organisms the most common ones are barnacles (infraclass *Cirripedia*), mussels, clams and oysters (class *Bivalvia*), tube-worms (family *Serpulidae*), sea-mats (phylum *Bryozoa*), hydroids (class *Hydrozoa*), and sea-squirts (class *Ascidiacea*) [5]. While all macrofouling organisms can be the cause of negative effects, calcareous macrofouling is held accountable for the most severe ones. According to Schultz [6], the increase in frictional resistance of biofouled surfaces escalates when calcareous macrofouling is involved. Although the settlement of marine organisms on submerged surfaces depends heavily on the conditions (temperature, salinity, etc.) of the aquatic environment [7], it is accepted that sessile barnacles (order *Sessilia*), commonly known as acorn barnacles, are among the most dominant calcareous biofouling organisms [8, 9].

Previous research on the effects of marine biofouling, either on the hydrodynamic properties of biofouled surfaces, or on the mechanical properties of antifouling coatings, has mainly been focused on experiments performed on plastic or metal plates that were exposed to a biofouling environment for long periods of time [6, 7, 10, 11]. However, such specimens are difficult to obtain, and the settlement of biofouling organisms cannot be easily monitored. The hydrodynamic properties of surfaces with barnacle fouling has been investigated both experimentally and computationally with artificially created surfaces where ordered arrays of barnacle shapes were placed on a flat surface [12, 13], or on an aerofoil element model [14]. However, in reality, barnacles follow an irregular distribution and tend to form clusters. More recently, Demirel et al. [15] performed experiments in a towing tank, using metal plates that were progressively covered with square 3D printed tiles of barnacle clusters. Although the placement of tiles did not follow a typical ordered arrangement, the placement of tiles fully covered by barnacles next to unaffected smooth tiles is not representative of typical barnacle settlement behaviour. In the current work, an algorithm for the generation of more realistic barnacle-type surface roughness is presented. The algorithm mimics the barnacle settlement behaviour to generate rough surfaces covered by barnacle clusters and allows to create surfaces for a range of fouling states from the onset of fouling to full coverage. The generated surfaces can be used in CFD applications or 3D-printed for use in wind-tunnel or towing tank experiments.

2 Generation of Rough Surfaces With Barnacle Shaped Roughness Elements

Barnacles live their early lives as free swimming larvae (*Nauplii*), that mostly feed on phytoplankton to gather resources, until they metamorphose to *Cyprid* larvae [16]. During their cyprid stage they stop feeding and search for a suitable place to settle. The metamorphosis to a sessile barnacle will confine it to the same spot for the rest of its life, so a place that will provide food, favourable flow conditions, and opportunities for reproduction is required. While there are many factors that govern the barnacle cyprid settlement behaviour, such as surface morphology, existence of light sources and velocity gradient of water current [17, 18], researchers agree [16, 17, 19, 20] that cyprids follow a gregarious behaviour, preferring to settle in places near other already settled barnacles of their own species.

For an algorithm that mimics the barnacle settlement behaviour, the gregarious nature of barnacles had to be captured, as well as the territorial behaviour of cyprids, that would not allow a barnacle shape to be placed within another. Other factors such as light sources, or the initial surface morphology are not included in the current version of the algorithm.

2.1 Definition of the Barnacle Roughness Element

Barnacles reach adulthood when cyprid larvae attach themselves to an underwater surface and become sessile by creating their calcareous outer shell. According to Rainbow [16], sessile barnacle shells consist of a number of interlocking plates ranging from a single plate (e.g *Megatrema anglicum*) to eight (e.g. species of the genus *Catomerus*), based on the evolutionary path that the sessile barnacle species has followed. The shells of most common barnacle species (e.g. *Balanus glandula*, *Semibalanus balanoides*, *Cthamalus stellatus*, *Amphibalanus amphitrite*) consist of six interlocking plates that form a conical frustum-like structure. The barnacle's crustacean body is enclosed inside the shell's mantle cavity and is protected at the upper surface with the paired scuta and terga plates that can be opened and closed as trapdoors (Fig. 1 (left)).

While a living barnacle can move its pair of scuta and terga to feed when submerged, its shell remains static and its shape resembles a conical frustum. Sadique [21] used this kind of geometry to model barnacle shapes and examine the flow around single individuals, or arrays of the same geometry [13]. The barnacle shapes presented in his work were based on measurements of seven individuals of the *Amphibalanus amphitrite* species. He used their diameter ratio $\frac{2R_1}{2R_2}$, and their aspect ratio $\frac{H}{2R_1}$ to model one frustum with representative size and shape to the seven individuals.

The roughness elements generated in this work are conical frustums, as depicted in Fig. 1 (right), defined similarly to the ones used by Sadique [21]. Every barnacle shape is defined by its lower radius R_1, the upper and lower radii ratio $RR \equiv R_1/R_2$, and its aspect ratio $AR \equiv H/2R_1$. By varying these values, a population of sessile

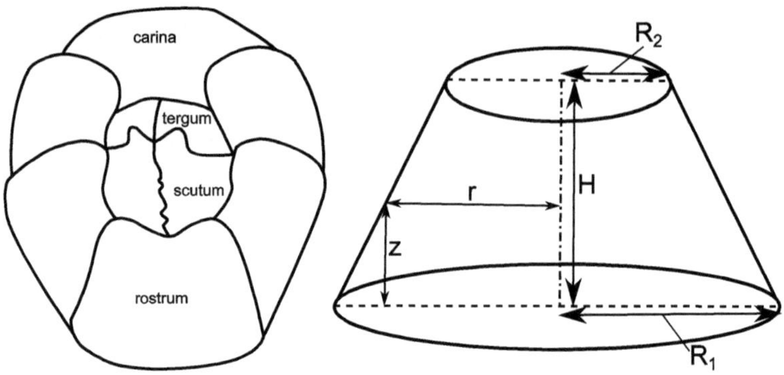

Fig. 1 Sketch of a barnacle with a six-plated shell structure (left), based on sketches of common barnacle species by Rainbow [16]. Frustum representation of barnacle shape (right)

barnacles with a realistic distribution of shape and size can be modelled. To this end, the values of ratios AR and RR are chosen randomly within user defined bounds, allowing for some variation in barnacle shapes, but still ensuring that no barnacle is unnaturally shaped. Given the actual values for the two radii and the frustum height, the height of the frustum at point with distance r from its centre can be calculated as in Eq. (1)

$$h = \begin{cases} H\dfrac{R_1 - r}{R_1 - R_2}, & R_2 < r \leq R_1 \\ H & , r \leq R_2. \end{cases} \tag{1}$$

2.2 *Barnacle Placement*

With the definition of the barnacle geometric characteristics a set of barnacle shapes for fouling of a surface can be modelled. The next task is the placement of barnacles on the smooth reference surface to create realistic barnacle colonies on the final rough surface. The location and occupied space of each barnacle can be defined by their lower radius, and the Cartesian coordinates (x_0, y_0) of their bases' centre, as illustrated in Fig. 2. In the following, we assume that barnacles are placed on a flat rectangular reference surface of size $x_{max} \times y_{max}$.

2.2.1 Generic and Seed Barnacles

As described in Sect. 2, the presented algorithm generates realistic barnacle colonies mainly by reproducing the gregarious behaviour of barnacles when choosing a location to settle. To this end, all barnacles are placed on the reference surface sequentially, while considering already placed barnacles as mature individuals. When a new barnacle is placed, one of the already placed barnacles is randomly chosen as its *neighbour*. The new barnacle will try to settle in the vicinity of this neighbour.

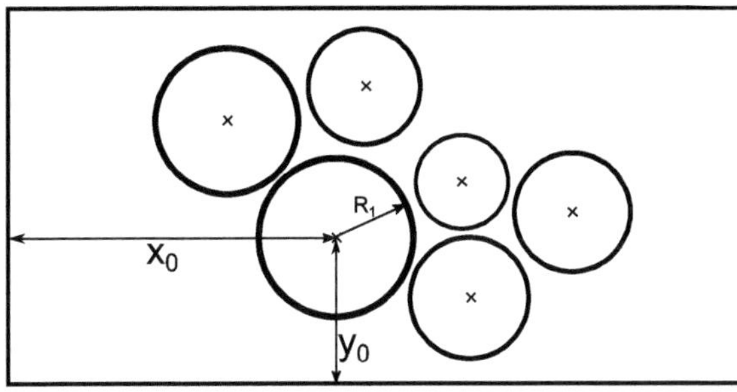

Fig. 2 The location of a barnacle is defined by the Cartesian coordinates of its base centre. The area a barnacle occupies is fully defined by its lower radius

The origin of the new barnacle is defined by distance d from the neighbour's centre and angle ϕ as illustrated in Fig. 3 (left). The Cartesian coordinates of the new barnacle's centre can be calculated as in Eq. (2), where subscript n denotes the neighbour barnacle origin coordinates. Both distance d, and angle ϕ are defined randomly. The bounds of d are controlled by the proximity factor, which is discussed in Sect. 2.2.2, and ϕ is defined within $[0, 2\pi)$. In case the origin of a new barnacle falls without the boundaries of the reference surface, the original choice for the neighbour is discarded and another barnacle is selected as its neighbour

$$\begin{aligned} x_0 &= x_{0_n} + d \cdot \cos\phi \\ y_0 &= y_{0_n} + d \cdot \sin\phi . \end{aligned} \tag{2}$$

The proposed method for placing new barnacle elements on the flat surface requires that at least one initial barnacle shape, i.e. a *seed* barnacle, should already be present. Consequently, the first barnacle will be placed randomly within the boundaries of the empty reference surface as starting point for a barnacle colony. The origin coordinates for the first barnacle are defined within the bounds shown in Eq. (3), to ensure that the whole body of the first barnacle will lie within the surface

$$\begin{aligned} x_0 &\in [R_1,\ x_{max} - R_1] \\ y_0 &\in [R_1,\ y_{max} - R_1] . \end{aligned} \tag{3}$$

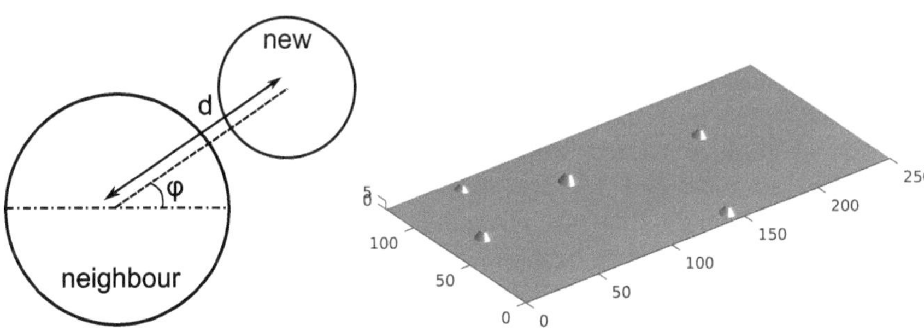

Fig. 3 Positioning a new generic barnacle with respect to its neighbour (left). Distribution of seed barnacles that will attract generic ones (right). Dimensions in mm

After the first barnacle is placed, the location of the rest of the barnacle shapes can be determined with the proposed methodology. However, if all these shapes are placed within the vicinity of the first barnacle or the ones following, only one barnacle colony will be formed. To enable the formation of several clusters, the number of seed barnacles can be increased. An example of seed barnacle arrangement is presented in Fig. 3 (right). When *generic* barnacles will be added to the surface, seed barnacles will attract them to create clusters.

2.2.2 Regulating Barnacle Proximity

While the barnacle placement method described in Sect. 2.2 enables the generation of barnacle colonies, with new barnacles settling within the vicinity of already settled mature barnacles, the new barnacle may overlap with other already settled barnacles that are not its selected neighbour. To avoid unwanted overlapping of barnacle shapes when the location of a new barnacle is randomly generated, its distance from all the existing barnacles is checked. The minimum allowed distance of the new barnacle from all existing barnacles is defined with the use of a proximity factor $f_p > 0$ that is used as shown in Eq. (4), where subscripts e and s denote any already existing barnacle and the new settling barnacle, respectively, while O is their origin

$$|\mathbf{O_e O_s}| \geq f_p \cdot (R_{1_e} + R_{1_s}). \tag{4}$$

When placing a barnacle, the distance d from its selected neighbour is set to a value $d > f_p(R_{1_n} + R_{1_s})$, so that the proximity criterion is fulfilled with respect to the selected neighbour. To ensure that the proximity criterion is also fulfilled with respect to all other existing barnacles, Eq. (4) is evaluated for these before settling the new one. If the proximity check fails for any settled barnacle, the original neighbour is discarded and a new neighbour is selected among the existing barnacles, and the proximity check is repeated. After a large number of failed attempts to settle, the new barnacle is not placed on the surface, and the algorithm terminates. This part of the settling methodology was inspired by the barnacle cyprid walk and territorial behaviour [16].

The value of f_p controls the density of barnacle clusters. With $0 < f_p < 1$, overlapping of barnacles is allowed and tightly packed barnacle colonies are generated. When $f_p = 1$ barnacles are allowed to touch, while with $f_p > 1$ no touching is allowed and loosely packed colonies are generated. The effect of f_p can be observed in Fig. 4, where three surfaces with the same number of barnacles, but different values of f_p are illustrated. The proximity check is used for the placement of all barnacles apart from the first one, although different values are used for seed and generic barnacles, so that seed barnacles can be more widespread resulting in more distinct barnacle clusters. Although random values of the proximity factor can also be used, in this work the proximity factors for seed and generic type barnacles are fixed for that specific barnacle type.

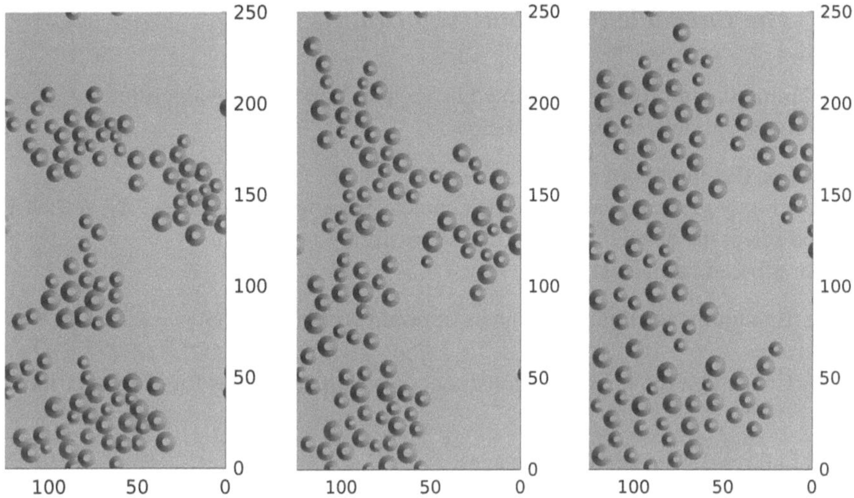

Fig. 4 Barnacle covered surfaces containing the same number of barnacles, generated with proximity factors $f_p = 0.8$ (left), $f_p = 1$ (middle), and $f_p = 1.1$ (right). Dimensions in mm

2.2.3 Resolving Periodicity

When generating rough surfaces for CFD simulations or for tiling of wind-tunnel floors it is often necessary for the surface to be periodic in one or both horizontal directions. Periodicity can be enforced in the current surface roughness generation algorithm by a simple modification: When a barnacle's origin is positioned at a distance less than its radius from the edges of the flat plate, the portion of its geometry that exceeds the surface boundaries must be copied to the opposite periodic position. In case the new barnacle is placed near one of the corners of the flat rectangular surface, portions of the barnacle need to be copied to the three other corners. Examples of periodic barnacles are illustrated in Fig. 5. When periodicity is included, proximity of a new barnacle to existing ones needs to be calculated taking into account the periodic boundaries.

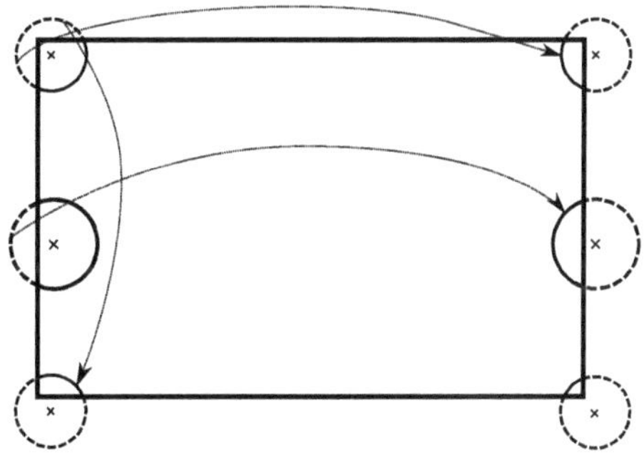

Fig. 5 Periodicity can be imposed by copying exceeding parts of barnacles to the opposite sides of the flat surface. Special care has to be taken when the barnacle is settled close to a corner of the surface and different parts have to be copied to different surface sides

2.3 The Barnacle Placement Algorithm

Combining the information presented in Sects. 2.1 and 2.2, the algorithm for generating rough surfaces is described below:

1. Define the number of *generic* and *seed* barnacles to be placed.
2. Randomly generate the geometric characteristics for all barnacles (R_1, RR, AR) and select the desired number of seed barnacles.
3. For all seed barnacles:

   a. Randomly generate coordinates of origin within the reference surface boundaries.
   b. Check that the proximity constraint (see Eq. (4)) applies for proximity factor f_p^{seed}.

4. For all remaining generic barnacles:

   a. Pick one of the already placed barnacles as *neighbour*.
   b. Randomly generate parameters d and ϕ (see Fig. 3), and calculate the new barnacle's origin coordinates.
   c. Check that the proximity constraint applies for the proximity factor $f_p^{generic}$. If the proximity constraint is not satisfied, go to step 4a.
   d. Apply periodicity if required.

5. The algorithm ends when all barnacles have been placed on the surface, or when the proximity check in step 4c fails too many times, which means that there is not enough space left to place another barnacle.

3 Application

The rough surfaces generated with the proposed algorithm can be used in various applications involving the investigation of fluid flows over surfaces fouled with barnacles, both experimentally and in CFD. As an example, the barnacle covered surface illustrated in Fig. 6 has been used as a rough wall in a DNS of turbulent channel flow to establish its effect on near-wall turbulence using an embedded boundary approach [22]. The rough surface consists of 275 barnacles, eight of which were seed barnacles. Generic barnacles were placed with a proximity factor $f_p^{generic} = 0.8$, while $f_p^{seed} = 4$ was used for seed ones. The initial size of the rectangular flat surface was $250 \times 125\,\text{mm}^2$, and the lower radius R_1 of all barnacles was randomly defined within $[3.5, 6]\,\text{mm}$. Ratios RR and AR were defined within $[0.31, 0.4]$ and $[0.34, 0.42]$, respectively, based on measurements by Sadique [21], and d was bounded within $[0.8, 1.2](R_{1_s} + R_{1_n})$. For use in the DNS, the rough surface was non-dimensionalised with the mean channel half-height δ as reference. The size of the computational domain was $2\pi\delta \times \pi\delta$ in the streamwise and spanwise directions, while the largest barnacle height was 0.1267δ. The simulation

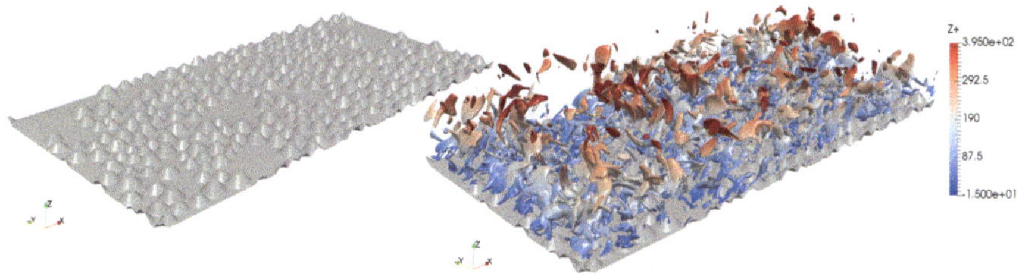

Fig. 6 Rough surface with 60% barnacle coverage (left) was used as boundary condition in DNS of turbulent channel flow. Instantaneous vortices are identified with the Q-criterion (Q = 15000) (right)

was performed at $Re_\tau = 395$ at a grid resolution of $864 \times 432 \times 512$ with uniform spacing in the streamwise and spanwise directions ($\Delta x^+ = \Delta y^+ = 2.8$). The spacing in the wall-normal direction is uniform up to the maximum roughness height with $\Delta z^+_{min} = 0.667$, and stretched above, reaching $\Delta z^+_{max} = 3.11$ at the channel centre.

The Hama roughness function ΔU^+ is defined as the downward shift of mean streamwise velocity profile compared to the equivalent of the flow over a smooth surface. The current simulation provided a roughness function value of $\Delta U^+ \approx 7.95$ characterizing the flow in the fully rough regime according to Nikuradse [23]. The equivalent sand-grain roughness of this surface can then approximated to be $k_{s,eqv} \approx 1.8k_{max}$, where k_{max} is the maximum peak-to-valley height of the rough surface, which coincides with the maximum barnacle height. Vortex identification with the Q-criterion (see Fig. 6 (right)) illustrates how the flow between the roughness elements can be investigated using this kind of simulations.

Although the proposed algorithm has been been introduced in the context of a flat rectangular reference surface, it can be implemented for other types of reference surfaces as well, such as curved surfaces within an hydrofoil design. Taking into account that the barnacle location and area is defined by its base origin and its lower radius, the barnacle bases could be projected on a curved surface allowing the remaining steps of the algorithm to proceed as described above.

4 Conclusion

In this work an algorithm for the generation of barnacle fouled rough surfaces has been presented. The proposed methodology mimics the settlement behaviour of barnacles, resulting in realistic examples of surfaces covered with barnacle clusters. The number of barnacle clusters can be controlled by adjusting the number of generic and seed barnacles that are placed on the surface, while the barnacle arrangement can be regulated by a proximity factor that controls how close new barnacles will be placed to already settled ones. As an example, a rough surface with 60% barnacle coverage has been used in DNS of rough-wall turbulent channel, giving the equivalent sand-grain roughness of the surface, and allowing a detailed investigation of changes in near-wall turbulence induced by this type of marine biofouling.

Acknowledgements This work was supported by the Engineering and Physical Sciences Research Council (EPSRC) [grant number EP/P009875/1].

References

1. Schultz MP, Bendick JA, Holm HR, Hertel WM (2011) Economic impact of barnacle fouling on a naval surface ship. Biofouling 27(1):87–89
2. Schultz MP (2007) Effects of coating roughness and biofouling on ship resistance and powering. Biofouling 23(5):331–341
3. Loxton J, Macleod AK, Nall CR, McCollin T, Machado I, Simas T, Vance T, Kenny C, Want A, Miller RG (2017) Setting an agenda for biofouling research for the marine renewable energy industry. Int J Mar Energy 19:292–303
4. Woods hole oceanographic institution (1952) Marine biofouling and its prevention. US Naval Institute, Annapolis, Maryland
5. Salta M, Chambers L, Wharton J, Wood R, Briand JF, Blache Y, Stokes KR (2009) Marine fouling organisms and their use in antifouling bioassays. In: Proceedings of the European corrosion congress (EUROCORR)
6. Schultz MP (2004) Frictional resistance of antifouling coating systems. J Fluids Eng 126:1039–1047
7. Vance TR, Fileman T (2014) ETI MA1001 - Reliable data acquisition platform for Tidal (ReDAPT) project: ME8.5 Final Report. Energy Technologies Institute, UK
8. Berntsson KM, Jonsson PR (2003) Temporal and spatial patterns in recruitment and succession of a temperate marine fouling assemblage: a comparison of static panels and boat hulls during the boating season. Biofouling 19(3):187–195
9. de Messano LVR, Sathler L, Reznik LY, Coutinho R (2009) The effect of biofouling on localized corrosion of the stainless steels N08904 and UNS S32760. Int Biodeterior Biodegrad 63:607–614
10. Andrewartha J, Perkins K, Sagrison J, Osborn J, Walker G, Henderson A, Hallegraeff G (2010) Drag forces and surface roughness measurments on freshwater biofouled surfaces. Biofouling 26(4):487–496
11. Lindholdt A, Dam-Johansen K, Olsen SM, Yebra DM, Kiil S (2015) Effects of biofouling development on drag forces of hull coatings for ocean-going ships: a review. J Coat Technol Res 12(3):415–444
12. Barros JM, Murphy EA, Schultz MP (2016) Particle image velocimetry measurements of the flow over barnacles in a turbulent boundary layer. In: Proceedings of the 18th international symposium on the application of laser and imaging techniques to fluid mechanics
13. Sadique J, Yang XI, Meneveau C, Mittal R (2015) Simulation of boundary layer flows over biofouled surfaces. In: Proceedings of the 22nd AIAA computational fluid dynamics conference (AIAA 2015-2616)
14. Orme JAC, Masters I, Griffiths RT (2001) Investigation of the effect of biofouling on the efficiency of marine current turbines. In: French C (ed) Proceedings of the MAREC 2001, pp 91–99
15. Demirel YK, Uzun D, Zhang Y, Fang H-C, Day AH, Turan O (2017) Effect of barnacle fouling on ship resistance and powering. Biofouling 33(10):819–834
16. Rainbow PS (1984) An introduction to the biology of British littoral barnacles. Field Stud 6:1–51
17. Crisp DJ (1955) The behaviour of barnacle cyprids in relation to water movement over a surface. J Exp Biol 32(3):569–590
18. Mullineaux LS, Butman CA (1991) Initial contact, exploration and attachment of barnacle (Balanus amphitrite) cyprids settling in flow. Mar Biol 110:93–103
19. Crisp DJ (1961) Territorial behaviour in barnacle settlement. J Exp Biol 38:429–446

20. Knight-Jones EW, Crisp DJ (1953) Gregariousness in barnacles in relation to the fouling of ships and to anti-fouling research. Nature 171:1109–1110
21. Sadique J (2016) Turbulent flows over macro-scale roughness elements - from biofouling barnacles to urban canopies. John Hopkins University, Baltimore
22. Busse A, Lützner M, Sandham ND (2015) Direct numerical simulation of turbulent flow over a rough surface based on a surface scan. Comput Fluids 116:129–147
23. Nikuradse J (1933) Strömungsgesetze in rauhen Rohren. VDI Forschungsheft 361

A Higher-Order Chimera Method Based on Moving Least Squares

Luis Ramírez, Xesús Nogueira, Pablo Ouro, Fermín Navarrina, Sofiane Khelladi and Ignasi Colominas

Abstract The Chimera/overset approach is widely used in the numerical simulation of flows involving moving bodies. In this approach, first used by Steger et al. in 1983, the domain is subdivided into a set of overlapping grids, which provide flexible grid adaptation, the ability to handle complex geometries and the relative motion of bodies in dynamic simulations. However, most of current methods present a second order convergence at most, due to the interpolation between overlapped grids. In this work a higher-order (>2) accurate finite volume method for the resolution of the Euler/Navier–Stokes equations on Chimera grids is presented. The formulation is based on the use of Moving Least Squares (MLS) approximations for transmission of information between the overlapped grids. The accuracy and performance of the proposed method is demonstrated by solving different benchmark problems.

1 Introduction

The development of high-accurate numerical method for the simulation of problems involving moving boundaries remains an active research field. The use of a single body-fitted mesh can require the deformation of it at every time step [1]. For simple movements, such as rotating sub-domains or sliding planes, it is possible to use the sliding mesh technique [2, 3]. Other techniques, such as the Immersed Boundary method or cut-cell techniques [4, 5] are attractive due its simplicity, although the

L. Ramírez (✉) · X. Nogueira · F. Navarrina · I. Colominas
Group of Numerical Methods in Engineering, Universidade da Coruña,
Campus de Elviña, 15071 A Coruña, Spain
e-mail: luis.ramirez@udc.es

P. Ouro
Hydro-environmental Research Centre, School of Engineering, Cardiff University,
12-14 The Parade, Cardiff CF24 3AA, UK

S. Khelladi
Arts et Métiers Paris Tech, DynFluid Lab, 151 Boulevard de l'Hôpital,
75013 Paris, France

E. Ferrer and A. Montlaur (eds.), *Recent Advances in CFD for Wind and Tidal Offshore Turbines*, Springer Tracts in Mechanical Engineering,
https://doi.org/10.1007/978-3-030-11887-7_7

lack of accuracy, the non-divergence-free velocities for incompressible flows and the difficulty to get high-order accuracy near the walls are the main drawbacks of these methods.

In this context, the use of the Chimera method, first introduced by Steger et al. in 1983 [6], is attractive for arbitrary body motions. In this technique, the domain is subdivided into a set of overlapped meshes, where the system of equations is solved on each grid and are connected through an interpolation process to exchange information. This interpolation is crucial to obtain a higher-order Chimera scheme. Several authors, see for example [7], have shown that if a lower order interpolation is used, the global accuracy of the scheme decreases. Based on the work developed for a high-order sliding mesh technique [3] a high-order Chimera approach has been developed [8]. The present approach may be considered as a generalization of the approach presented by the authors in [3] in the simulation of bodies under arbitrary motions, where a meshless technique is used to transfer the information between the overlapped grids.

2 Governing Equations

The Navier–Stokes equations, written in general form as a system of conservation laws, read as

$$\frac{\partial \boldsymbol{U}}{\partial t} + \nabla \cdot \left(\boldsymbol{F}^H - \boldsymbol{F}^V\right) = \boldsymbol{0} \tag{1}$$

where $\boldsymbol{U}$ is the vector of conservative variables, and the fluxes are split into a hyperbolic-like part, $\boldsymbol{F}^H$, and an elliptic-like part, $\boldsymbol{F}^V$.

For a given control volume I, the finite volume discretization of the system of conservation laws, (1) is

$$\Omega_I \frac{\partial \boldsymbol{U}_I}{\partial t} = \sum_{j=1}^{N_f} \sum_{ig=1}^{N_G} \left[\left(\boldsymbol{F}^V - \boldsymbol{F}^H\right) \cdot \hat{\boldsymbol{n}}_j\right]_{ig} \mathcal{W}_{ig} \tag{2}$$

where Ω_I is the area associated to the control volume I, $\boldsymbol{U}_I$ represents the mean value of the conservative variables $\boldsymbol{U}$ over the control volume I, N_f is the number of edges (in 2D) of the control volume, N_G represents the number of quadrature points for each edge, $\mathcal{W}_{ig}$ is the corresponding quadrature weight for the quadrature point (ig) at each cell edge, and $\hat{\boldsymbol{n}}_j$ is the unitary normal $\boldsymbol{n}$ times the length of edge j.

In order to achieve a high order finite volume method, there is a need for an accurate reconstruction of the variables inside each control volume. In this work, we use the MLS method to get the required accuracy.

3 MLS Method

The Moving Least Squares (MLS) approach was originally devised in 1981 for data processing and surface generation [9]. The basic idea of the MLS approach is to approximate a function $u(\mathbf{x})$ at a given point $\mathbf{x}$ through a weighted least squares fitting on a compact domain around $\mathbf{x}$. The resulting approximation, $\widehat{u}(\mathbf{x})$, can be written as

$$u(\mathbf{x}) \approx \widehat{u}(\mathbf{x}) = \mathbf{N}^T(\mathbf{x})\mathbf{u}_{\Omega_\mathbf{x}} = \mathbf{p}^T(\mathbf{x})\mathbf{M}^{-1}(\mathbf{x})\mathbf{P}_{\Omega_\mathbf{x}}\mathbf{W}(\mathbf{x})\mathbf{u}_{\Omega_\mathbf{x}} \tag{3}$$

where $\mathbf{N}^T(\mathbf{x})$ is the vector of MLS shape functions and $\mathbf{u}_{\Omega_\mathbf{x}}$ contains the known values of the function $u(\mathbf{x})$ at the compact domain $\Omega_\mathbf{x}$. The discretized domain, namely stencil, is formed with the cell centroids of the neighboring cells. The size of $\mathbf{N}^T(\mathbf{x})$ is equal to the number of neighboring cells from the stencil. The vector $\mathbf{p}^T(\mathbf{x})$ represents an m-dimensional polynomial basis, $\mathbf{P}_{\Omega_\mathbf{x}}$ is defined as a matrix where the basis functions are evaluated at each point of the stencil and the moment matrix $\mathbf{M}(\mathbf{x})$ is obtained through the minimization of an error functional [10, 11], and is defined as

$$\mathbf{M}(\mathbf{x}) = \mathbf{P}_{\Omega_\mathbf{x}}\mathbf{W}(\mathbf{x})\mathbf{P}^T_{\Omega_\mathbf{x}} \tag{4}$$

The *kernel* or smoothing function, $W(\mathbf{x})$, weights the importance of the different points that take place in the approximation. Among the wide variety of *kernel* functions [12], we have used the exponential *kernel*, defined as

$$W_j(x_j, x, s_x) = \frac{e^{-\left(\frac{d}{c}\right)^2} - e^{-\left(\frac{d_m}{c}\right)^2}}{1 - e^{-\left(\frac{d_m}{c}\right)^2}} \tag{5}$$

for $j = 1, \ldots, n_x$, where $d = \left|x_j - x\right|$, $d_m = 2\max\left(\left|x_j - x\right|\right)$.

In Eq. 5, d_m is the *smoothing length*, n_x is the number of neighbors and x is the reference point where the compact support is centered, the coefficient c is defined as $c = \frac{d_m}{s_x}$ and s_x is the shape parameter of the *kernel*. This parameter defines the *kernel* properties and therefore, the properties of the numerical scheme obtained.

The order of MLS approximations is determined by the polynomial basis used in the construction of MLS shape functions. In this work we use a quadratic polynomial basis. The high-order approximate derivatives of the field variables $u(\mathbf{x})$ can be expressed in terms of the derivatives of the MLS shape function. Hence, the nth derivative is defined as

$$\frac{\partial^n \widehat{u}}{\partial x^n} = \sum_{j=1}^{n_x} \frac{\partial^n N_j(\mathbf{x})}{\partial x^n} u_j \tag{6}$$

We refer the interested reader to [10, 13, 14] for a complete description of the computation of MLS derivatives.

4 The Chimera Method

The high-order Chimera method was implemented on a high-order finite volume method, the FV-MLS, which adopts MLS approximations. For further details of the FV-MLS method the reader is referred to [15]. The basic steps of the Chimera method are:

1. **Identify the resolved cells**: we identify the cells where the system of conservation laws is computed. We label them as **non-overlapped** cells. This set is comprised of all the cells of the near-body grid and those cells of the background grid that are not **completely** covered by the near-body grid and the solid body. A hole-cutting process is used to determine the overlapped cells on the background grid.
2. **Solve the system of equations**: the governing equations are solved for each grid independently. During the hole-cutting process, two different new interfaces are identified. Since we are using a finite volume method, there is a need to identify the right and left states on both sides of these interfaces in order to compute the fluxes. We accomplish this creating a fictitious cell to define completely define the fluxes. This is schematically shown in Fig. 1. Following [3], the mean value of the variables at the fictitious cell are computed using Moving Least Squares as follows

$$\boldsymbol{U}_{Halo} = \frac{1}{A_{Halo}} \int \boldsymbol{U} dA = \frac{1}{A_{Halo}} \int \sum_{j=1}^{n_x} N_j(\boldsymbol{x}_{Halo}) U_j dA \tag{7}$$

   where A_{Halo} is the area associated to the halo cell I_{halo} and $\boldsymbol{N}(\boldsymbol{x}_{Halo})$ is the vector of MLS shape functions centered at the centroid of I_{halo}.

We refer the reader to [8] for a further details on the proposed high-order Chimera method.

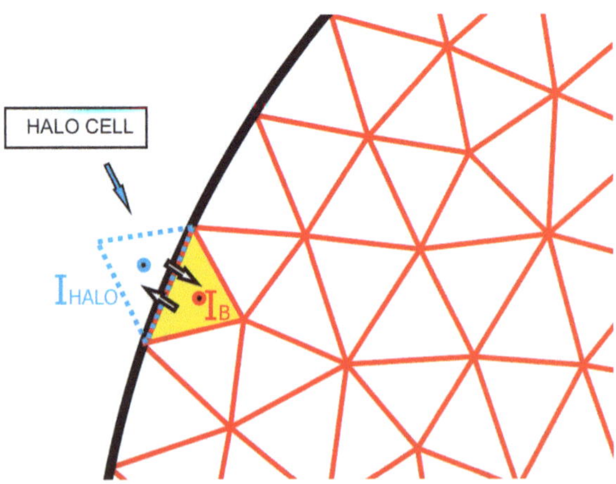

Fig. 1 Schematic representation of the flux exchange at the interfaces between overlapped grids

5 Numerical Examples

This section presents the numerical results for several benchmark problems defined with the aim of assessing the accuracy of the proposed methodology.

5.1 *Isentropic Vortex Convection*

The first validation test case corresponds to the unsteady vortex convection. This test case is widely used as benchmark for moving grid formulations [3, 16–19], since it has analytical solution, that reads

$$\begin{aligned}
\frac{u(x,y,t)}{a_\infty} &= \frac{u_\infty}{a_\infty} - \frac{K}{2\pi a_\infty}\hat{y}e^{\alpha(1-r^2)/2} \\
\frac{v(x,y,t)}{a_\infty} &= \frac{v_\infty}{a_\infty} + \frac{K}{2\pi a_\infty}\hat{x}e^{\alpha(1-r^2)/2} \\
\frac{T(x,y,t)}{T_\infty} &= 1 - \frac{K^2(\gamma-1)}{8\alpha\pi^2 a_\infty^2}e^{\alpha(1-r^2)} \\
\frac{\rho(x,y,t)}{\rho_\infty} &= \left(\frac{T(x,y,t)}{T_\infty}\right)^{\frac{1}{\gamma-1}} \\
\frac{p(x,y,t)}{p_\infty} &= \left(\frac{T(x,y,t)}{T_\infty}\right)^{\frac{\gamma}{\gamma-1}}
\end{aligned}$$

where $\hat{x} = x - x_0 - u_\infty t$, $\hat{y} = y - y_0 - v_\infty t$ and $r = \sqrt{\hat{x}^2 + \hat{y}^2}$. Here, the chosen parameters are $\alpha = 1$, $\rho_\infty = 1$, $p_\infty = 1$, $(u_\infty, v_\infty) = (2, 2)$, $(x_0, y_0) = (-5, -5)$ and $K = 5$. With this set of parameters the vortex starts at the position $(x, y) = (-5, -5)$ and at $t = 5$ reaches the final position $(x, y) = (5, 5)$. The position of the overlapped grid varies with time according to $(x, y)^T = (u_\infty t, v_\infty t)^T$. In Fig. 2 the initial condition and the initial mesh configuration are shown.

The L_2 norm obtained with a 3rd order FV-MLS methods is shown in Fig. 3. It is observed that the formal order of accuracy is recovered.

5.2 *Flow Around a Fixed Cylinder*

The second validation test corresponds to the unsteady flow around a cylinder at $Re = 100$. The freestream Mach number is $M_\infty = 0.1$ and the diameter of the cylinder is $D = 1$. We impose the no-slip and adiabatic boundary conditions at solid walls. The computational domain is discretized using two grids, A and B, with 10114 and 3600 cells respectively, and 200 elements along the cylinder surface. In Table 1,

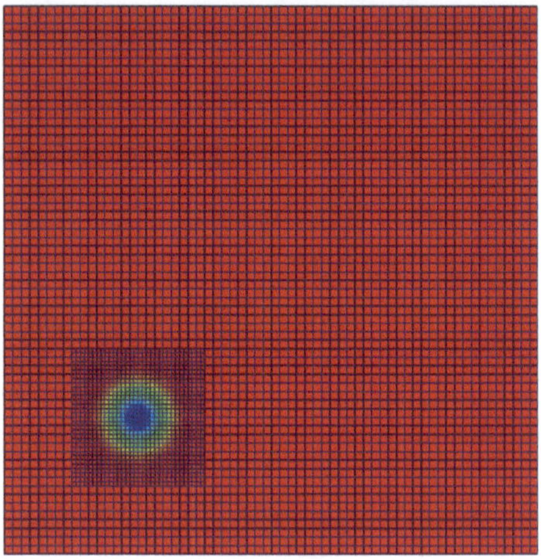

Fig. 2 2D Isentropic vortex convection: initial condition

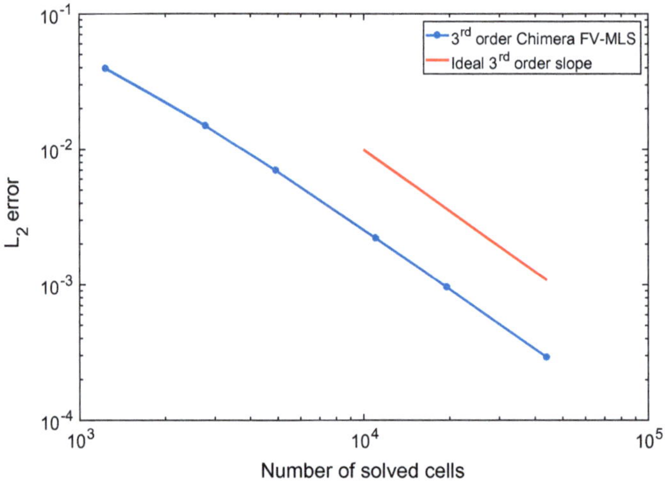

Fig. 3 2D Isentropic vortex convection: L_2 norm converge rate

we compare the obtained parameters (mean drag coefficient $\overline{C}_D$, lift amplitude C_L' and Strouhal number St) with the ones obtained with other methods [20, 21]. The parameters obtained with the new FV-MLS Chimera method agree well with those obtained by other authors.

5.3 Flow Past an Oscillating Circular Cylinder

In this last numerical example, we compute the flow around an oscillating cylinder. The aim of this test case is to analyze a cross-flow moving boundary, in order to

Table 1 Flow past a fixed circular cylinder at $Re = 100$. The time-average drag coefficient $\overline{C}_D$, the amplitude of the lift coefficient C_L' and the Strouhal number St

Method	$\overline{C}_D$	C_L'	St
Present method	1.361	0.331	0.165
Yang et al. [20]	1.393	0.335	0.165
Kara et al. [21]	1.360	0.340	0.164

validate the Chimera method when moving boundaries are present. The location of the center of the cylinder, $(x_c, y_c)^T$, is imposed as $x_c = 0$ and $y_c = A\sin(2\pi f_e t)$, where A is the amplitude of oscillation and (f_e) is the oscillation frequency. Following [22], we analyze an amplitude of $A = 0.2$ and a set of frequencies $\frac{f_e}{f_0} = 0.8, 0.9, 1.0, 1.1, 1.2$, where the quantity f_0 denotes the natural frequency of vortex shedding for a static test case, which is previously computed from a static configuration. The mesh resolution is the same as the one employed in Sect. 5.2.

In Fig. 4, we show the root mean square values of the drag and lift coefficients (C_{Drms}, C_{Lrms}). It is observed a good agreement between our results and those obtained by Guilmineau et al. [22] with a mesh resolution four times finer.

Finally, the time evolution of the drag and lift coefficient obtained for the cases $f_e = 0.9 f_0$ and $f_e = 1.2 f_0$ are plotted in Fig. 5.

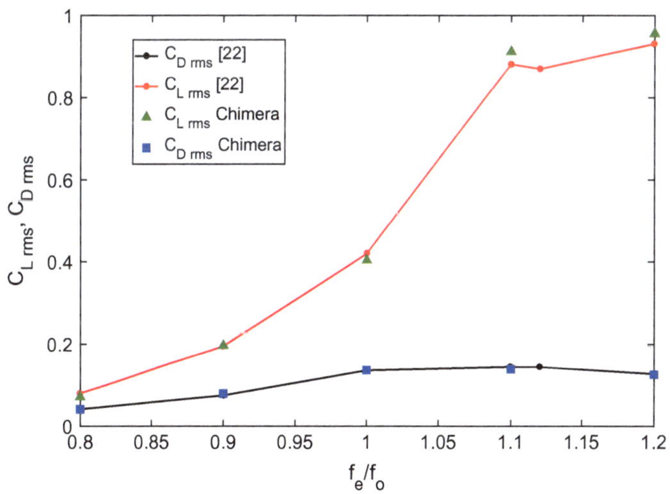

Fig. 4 Flow past an oscillating circular cylinder: root mean square values of drag and lift coefficients (C_{Drms} and C_{Lrms}) versus oscillating frequency f_e/f_0

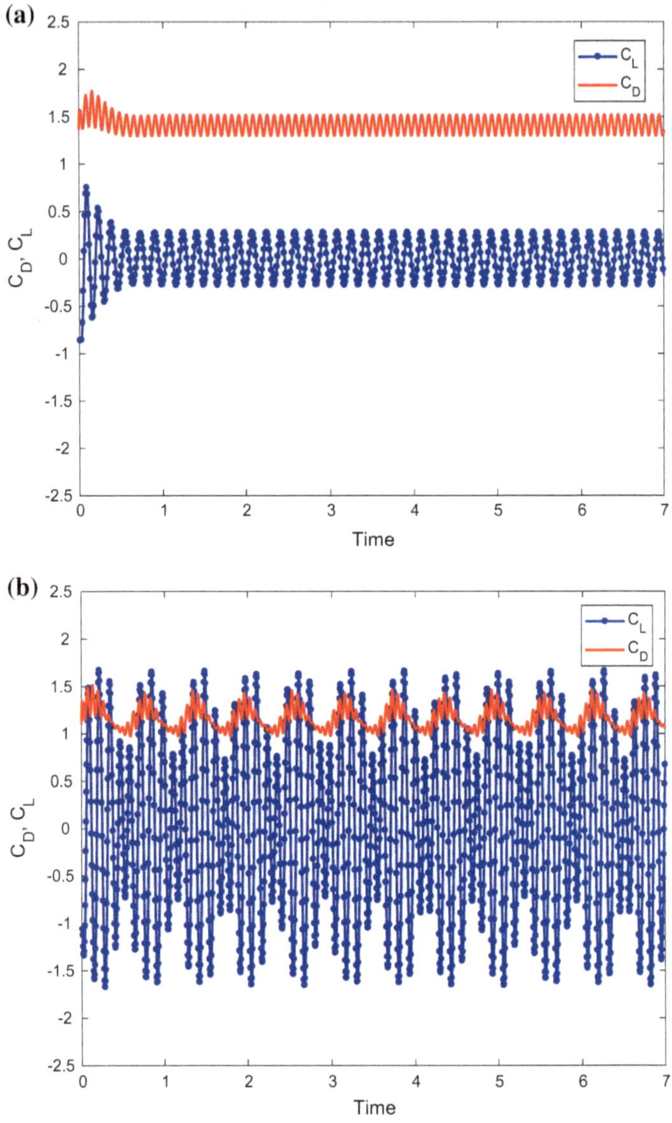

Fig. 5 Flow past an oscillating circular cylinder: prescribed motion with: **a** $f_e = 0.9 f_0$ and **b** $f_e = 1.2 f_0$. Drag and lift coefficients (C_D and C_L) versus time

6 Conclusions

In this work, we have presented a new higher-order accurate Chimera method for overlapped arbitrary grids. The method is based on Moving Least Squares approximations, which transfer accurately the information between grids. The new method is implemented on a high-order finite volume scheme, and the numerical results have shown that the order of the scheme is preserved. Moreover, we have shown that the results obtained with the new method are comparable to those obtained using other methods previously published using finer grid resolution.

References

1. Boscheri W, Dumbser M (2013) Arbitrary-Lagrangian–Eulerian one-step WENO finite volume schemes on unstructured triangular meshes. Commun Comput Phys 14:1174–1206
2. Ferrer E, Willden RHJ (2012) A high order discontinuous Galerkin - Fourier incompressible 3D Navier–Stokes solver with rotating sliding meshes. J Comput Phys 231:7037–7056
3. Ramirez L, Foulquié C, Nogueira X, Khelladi S, Chassaing JC, Colominas I (2015) New high-resolution-preserving sliding mesh techniques for higher order finite volume schemes. Comput Fluids 118:114–130
4. Peskin CS (1972) Flow patterns around heart valves: a numerical method. J Comput Phys 10:252–271
5. Ouro P, Cea L, Ramirez L, Nogueira X (2016) An immersed boundary method for unstructured meshes in depth averaged shallow water models. Int J Numer Methods Fluids 81:672–688
6. Steger J, Dougherty F, Benek J (1982) A Chimera grid scheme. In: ASME mini-symposium on advances in grid generation, Houston
7. Delfs JW (2001) An overlapped grid technique for high resolution CAA schemes for complex geometries. AIAA paper 2001–2199
8. Ramirez L, Nogueira X, Ouro P, Navarrina F, Khelladi S, Colominas I (2018) A higher-order Chimera method for finite volume schemes. Arch Comput Methods Eng 25:691–706
9. Lancaster P, Salkauskas K (1981) Surfaces generated by moving least squares methods. Math Comput 37:141–158
10. Cueto-Felgueroso L, Colominas I, Nogueira X, Navarrina F, Casteleiro M (2007) Finite-volume solvers and moving least squares approximations for the compressible Navier–Stokes equations on unstructured grids. Comput Methods Appl Mech Eng 196:4712–4736
11. Liu WK, Hao W, Chen Y, Jun S, Gosz J (1997) Multiresolution reproducing kernel particle methods. Comput Mech 20:295–309
12. Liu GR, Liu MB (2003) Smoothed particle hydrodynamics. A meshfree particle method. World Scientific Publishing, Singapore
13. Chassaing JC, Khelladi S, Nogueira X (2013) Accuracy assessment of a high-order moving least squares finite volume method for compressible flows. Comput Fluids 71:41–53
14. Khelladi S, Nogueira X, Bakir F, Colominas I (2011) Toward a higher order unsteady finite volume solver based on reproducing kernel methods. Comput Methods Appl Mech Eng 200:2348–2362
15. Nogueira X, Ramirez L, Khelladi S, Chassaing JC, Colominas I (2016) A high-order density-based finite volume method for the computation of all-speed flows. Comput Methods Appl Mech Eng 298:229–251
16. Lee KR, Park JH, Kim KH (2011) High-order interpolation method for overset grid based on finite volume method. AIAA J 49:1387–1398
17. Sherer SE, Scott JN (2005) High-order compact finite-difference methods on general overset grids. J Comput Phys 210:459–496
18. Wang G, Duchaine F, Papadogiannis D, Duran I, Moreau S, Gicquel LYM (2014) An overset grid method for large eddy simulation of turbomachinery stages. J Comput Phys 274:343–355
19. Galbraith MC, Benek JA, Orkwis PD, Turner MG (2014) A discontinuous Galerkin Chimera scheme. Comput Fluids 98:27–53

20. Yang X, Zhang X, Li Z, He GW (2009) A smoothing technique for discrete delta functions with application to immersed boundary method in moving boundary simulations. J Comput Phys 228:7821–7836
21. Kara MC, Stoesser T, McSherry R (2015) Calculation of fluid-structure interaction: methods, refinements, applications. Proc ICE Eng Comput Mech 168(2):59–78
22. Guilmineau E, Queutey P (2002) A numerical simulation of vortex shedding from an oscillating circular cylinder. J Fluids Struct 16:773–794

A Review on Two Methods to Detect Spatio-Temporal Patterns in Wind Turbines

Soledad Le Clainche, José M. Vega, Xuerui Mao and Esteban Ferrer

Abstract This Chapter presents a review on two methods for the analysis of flow structures in wind turbines. These methods are higher order dynamic mode decomposition and spatio-temporal Koopman decomposition, which are highly efficient tools suitable for the detection of spatio-temporal patterns in complex flows. These two techniques have been applied to detect the main flow structures in a cross-flow wind turbine in turbulent regime, and in an horizontal wind turbine, which is laminar in the near field but transitioning to turbulence in the far field. Using these methods, a reduced number of traveling waves which are responsible for triggering the flow transition, are able to describe the aforementioned complex flows.

1 Introduction

The use of renewable energies is a topic that has recovered special interest during the last few years. The recent awareness of environmental problems, such as the hole in the ozone layer or the green-house effect has led to the strong need of replacing fuels and oils by alternative sources of energies. Wind energy is an environmentally friendly source of energy that has become very popular during the last few years [1]. However, a strong effort is still necessary to maximize the energy extraction with wind turbines.

Wake interactions in wind farms may lead to highly complex non-linear phenomena that limit the complete description of the flow physics on wind turbines. Despite

S. Le Clainche (✉) · J. M. Vega · E. Ferrer
School of Aerospace Engineering, Universidad Politécnica de Madrid, 28040 Madrid, Spain
e-mail: soledad.leclainche@upm.es

J. M. Vega
e-mail: josemanuel.vega@upm.es

E. Ferrer
e-mail: esteban.ferrer@upm.es

X. Mao
Faculty of Engineering, The University of Nottingham, Nottingham NG7 2RD, UK
e-mail: Xuerui.Mao@nottingham.ac.uk

E. Ferrer and A. Montlaur (eds.), *Recent Advances in CFD for Wind and Tidal Offshore Turbines*, Springer Tracts in Mechanical Engineering,
https://doi.org/10.1007/978-3-030-11887-7_8

of its complexity, the analysis of flow structures in wind turbines is highly relevant in order to extract significant information to improve the performance of wind farms and to maximize the generation of power. If the physical phenomena describing a problem is understood, then it is possible to modify the flow (flow control). This Chapter presents a review on different ways to detect flow patterns in wind turbines. The spatio-temporal structures describing the flow of two types of wind turbines, cross-flow [2, 3] and horizontal axis [4], will be presented using (i) spatio-temporal diagrams and (ii) two highly efficient tools, which are based on the well-known technique dynamic mode decomposition [5]. These are higher order dynamic mode decomposition (HODMD) [6] and spatio-temporal Koopman decomposition (STKD) [7]. These techniques represent the evolution of spatio-temporal flow structures as either a group of modes that oscillate in time (HODMD) or as a group of traveling waves that move with a certain phase velocity (STKD). In this way, it is possible to simplify very complex flow features, as the ones found in the wake of a wind turbine, to understand flow behavior. In this Chapter, HODMD and STKD are applied to detect spatio-temporal patterns in wind turbines working in conditions of transitional and turbulent regimes.

The Chapter is organized as follows. Section 2 briefly introduces the method HODMD, while Sect. 3 applies the method to detect flow structures in a cross-flow wind turbine. Section 4 introduces the method STKD and it is applied in Sect. 5 to extract flow structures in a horizontal wind turbine. Finally, Sect. 6 presents the conclusions of this work.

2 Higher Order Dynamic Mode Decomposition

Higher order dynamic mode decomposition (HODMD) [6] is an extension of dynamic mode decomposition (DMD) [5] that has been recently introduced for the analysis of complex flows, such as noisy experiments [8], quasi-periodic solutions (exhibiting a large number of frequencies) or transitional-turbulent flows [9, 10].

As DMD, HODMD decomposes the original data $\mathbf{v}_k$ (snapshot) as an expansion of DMD modes in the following way

$$\mathbf{v}_k \simeq \mathbf{v}_k^{DMD} \equiv \sum_{m=1}^{M} a_m \mathbf{u}_m e^{(\delta_m + i\omega_m)(k-1)\Delta t}, \quad k = 1, \ldots, K, \tag{1}$$

where $\mathbf{u}_m$ are the DMD modes, and ω_m, δ_m, a_m are the associated frequencies, growth rates and amplitudes, respectively.

The algorithm of HODMD considers only two main steps. In the first step, a reduction of the spatial redundancies is carried out by means of either standard singular value decomposition (SVD) [11] or higher order singular value decomposition (HOSVD) [12]. This step can also be applied for noise filtering, if needed. A tolerance ε, set by the user, determines the number of SVD modes retained. The second step uses a DMD-like algorithm that considers a set of data that contains d time-delayed snapshots [13]. This treatment is similar to the well-known sliding window process in power spectral density (PSD) analysis. The latter step is called DMD-d and the parameter d, set by the user, can be compared to the number of segments used in PSD. A final tolerance ε_1, set by the user, determines the number of M DMD modes retained to form expansion (1). A deeper insight of this method can be found in previous work by the authors [6, 8].

3 HODMD for Cross-Flow Wind Turbines

Three-dimensional numerical simulations have been performed to study the wake of a three-bladed cross-flow turbine. The incompressible Navier–Stokes equations have been solved using a discontinuous Galerkin method with sliding meshes (more details in [14, 15]). The wind turbine is modeled using a sliding-mesh with three equispaced foils azimuthally spaced by 120°. Three-dimensional effects are taken into account with an expansion in the spanwise direction, using Fourier modes.

The boundary conditions used at inlet, walls and sides of the domain are of Dirichlet type for velocity and of vorticity type for pressure [16]. At the outflow Dirichlet boundary conditions are used for pressure and Neumann for velocity. High order no-slip conditions are used to model the surface of the three foils [16]. The number of elements used for the spatial discretization are 3676 elements, with polynomials of degree P $= 3$ in the plane parallel to the incoming flow and 16 planes in the spanwise direction, providing a total of 0.6 million degrees of freedom.

HODMD has been applied to study a cross-flow wind turbine, modeled with a sliding mesh and containing three foils, in turbulent regime, with Reynolds number $\mathrm{Re} = 1.5 \cdot 10^4$ defined as $\mathrm{Re} = U_t D/\nu$, where U_t is the free-stream velocity, D is the diameter of the wind turbine blades, and ν is the kinematic viscosity. The rotation velocity of the wind turbine is $\omega = 0.11$. Due to the high complexity of these data, two different types of analyses have been carried out. In the first analysis, HODMD has been applied to study the wake of the wind turbine. In the second analysis, the method has been applied to study the region of the turbine blades. A new strategy has been proposed in order to improve this second analysis. Figure 1 shows the instantaneous flow field of the velocity fluctuations computed in the wind turbine. Figure 1 middle shows the wake of the turbine. As seen, the flow is slightly asymmetric, possibly due to turbulent effects. Figure 1 right shows the spatio-temporal diagram of such velocity fluctuations in the wake of the turbine. In the figure, it is possible to distinguish a traveling wave describing the wake with phase velocity (slope) $1/c = \Delta t/\Delta x \simeq 17$, although this pattern is not homogeneous (positive and negative velocity fluctuations do not present the same shape), meaning that the asymmetry previously depicted raises due to temporal effects. The first HODMD analysis has been carried out to identify the main temporal patterns related with such asymmetry. Figure 1 left shows the streamwise velocity in the area of the wind turbine where the second HODMD

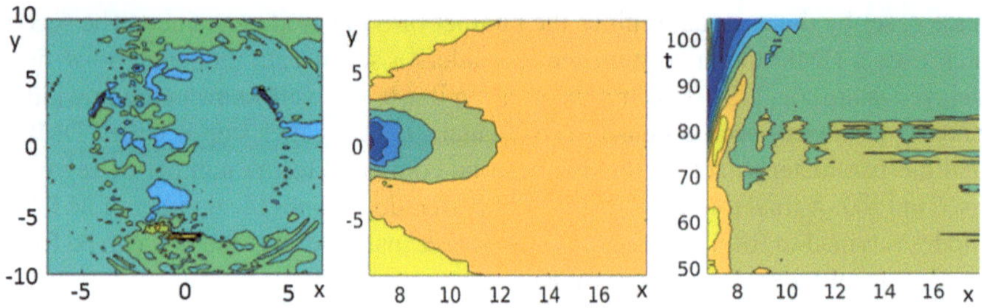

Fig. 1 Left: streamwise velocity in the area close to the turbine. Middle: streamwise velocity fluctuations in the wake of the wind turbine. Right: spatio-temporal diagram (time versus streamwise direction) of the streamwise velocity in the wake of the wind turbine at $y/D = 0.5$. Blue, green and yellow colors scale as -1, 0 and 1, respectively

analysis has been carried out. As seen, the complexity of the flow is larger in this region than in the wake, since it is possible to find a larger number of small size structures. HODMD analysis has been carried out in order to study the main structures interacting in this area, which are in charge of describing such complex flow.

First, HODMD has been applied to study the wake of the wind turbine using a set of 196 snapshots equispaced in time, with time step $\Delta t = 0.288$ time units. After some calibration (based on the robustness of the results [17]), the parameters used for HODMD analysis are $d = 12$, $\varepsilon = 0.005$ and $\varepsilon_1 = 0.05$. HODMD retains 12 modes (6 different frequencies plus their conjugate complex). The error made when these modes are used to reconstruct the original data using expansion (1) is RRMSE $\simeq 0.054$, meaning that the DMD approximation obtained is reasonably good. The reconstruction error obtained using the same parameters and $d = 1$ is RRMSE $\simeq 0.197$, justifying the need of using HODMD in this complex turbulent flow. Figure 2 left shows the frequencies versus amplitudes of the DMD modes. As seen, the modes decay spectrally. Although the regime of the flow analyzed is turbulent, the solution is periodic, with dominant frequency $\omega \simeq 0.11$, which is the frequency of rotation of the wind turbine. A sub-harmonic of this frequency, $\omega \simeq 0.07$, is also found in these calculations.

As second analysis, HODMD has been carried out in the region of the wind turbine. The number of snapshots used are the same as before, and the parameters used in the analysis are $d = 5$, $\varepsilon = 0.005$ and $\varepsilon_1 = 0.05$. The method retains 20 modes (10 different frequencies plus their conjugate complex). The reconstruction error of the three-dimensional original data is RRMSE $\simeq 0.25$, which is larger than in previous cases. The reason is that the complexity of the flow analyzed is larger, since the body is also moving in space (the turbine is rotating), justifying this error. Figure 2 right shows the frequencies versus amplitudes of the DMD modes. As in the previous case, the frequencies decay spectrally, but it is possible to distinguish two different branches. The first branch is conformed by the frequency of rotation of the three turbine blades, $\omega_3 \simeq 0.33$ (third harmonic of the rotation velocity of the wind turbine $\omega_1 \simeq 0.11$) and its harmonics (i.e.: $\omega_6 = 2\omega_3 = 6\omega_1$, etc.). The second branch (external) is conformed by the rotation frequency of the turbine, $\omega_1 \simeq 0.11$ (leading frequency) and the remaining harmonics of ω_1 which are no contained in the first branch ($\omega_4 = 4\omega_1, \omega_5 = 5\omega_1, \dots$). Table 1 details the frequencies calculated in each DMD analysis, presented in Fig. 2, where it is possible to see that all the modes, but the sub-harmonic of $\omega \simeq 0.11$, are found in both parts of the domain. The DMD modes explain the differences between the two branches of modes describing the wind turbine. Figure 3 shows the DMD modes obtained for ω_1 and ω_3 in both, the wind turbine and the wake. In the wake of the wind turbine, the shape of modes ω_1 and ω_3 is very similar, although the size of the spatial structures is smaller in the mode with highest frequency. On the contrary, in the area of the turbine these modes present different behaviour. The region of highest intensity in the mode with fundamental frequency ω_1 (branch 2) is located in the inner region of the blades of the wind turbine. The same effect is found in the remaining modes identified in branch 2. However, the modes located in branch 1, whose frequency is related to the rotation velocity of the three blades of the wind turbine ($\omega_3 \simeq 3\omega_1$), present a well defined structure in the region where the three blades are located. The same structures are found in the remaining modes of branch 1, but, as expected their spatial structure is smaller, since the frequency is larger. More details about these results can be found in [3]. Finally, a new analysis has been carried out in the region of the wind turbine aiming to study the spatio-temporal structures found in this region. To carry out this analysis, the Cartesian coordinate system x and y has been changed by polar coordinates as $x' = R\cos\theta$ and $y' = R\sin\theta$, with $R = 4$ the radius of the wind turbine. Figure 4 left shows the new shape of the flow field. The same HODMD analysis has been carried out, obtaining the same frequencies as before (shown in Table 1), but with higher accuracy (better approximation of the harmonics). From this coordinate change, it is possible to easily calculate the leading wave number along the new normal component y' applying HODMD along this component. This leading wavenumber is $\kappa' \simeq 0.13$, transformed to the original system leads to $\kappa \simeq 0.52$.

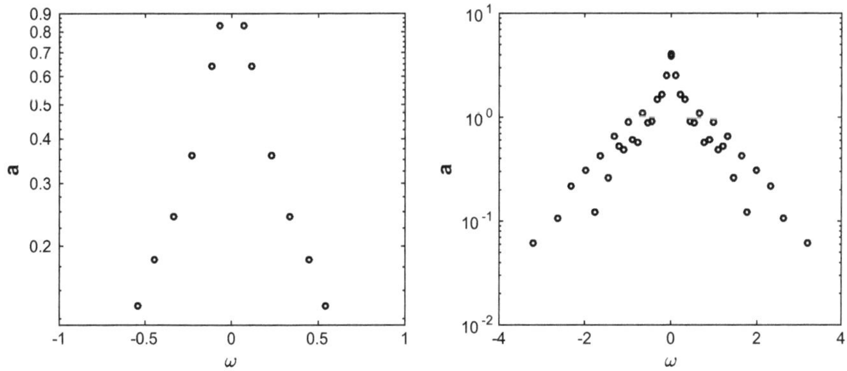

Fig. 2 Frequencies (ω_m) versus amplitudes (a_m) obtained in the HODMD analysis of the three-dimensional wake of the wind turbine. Left: analysis carried out in the wake of the wind turbine. Right: analysis carried out in the area close to the wind turbine

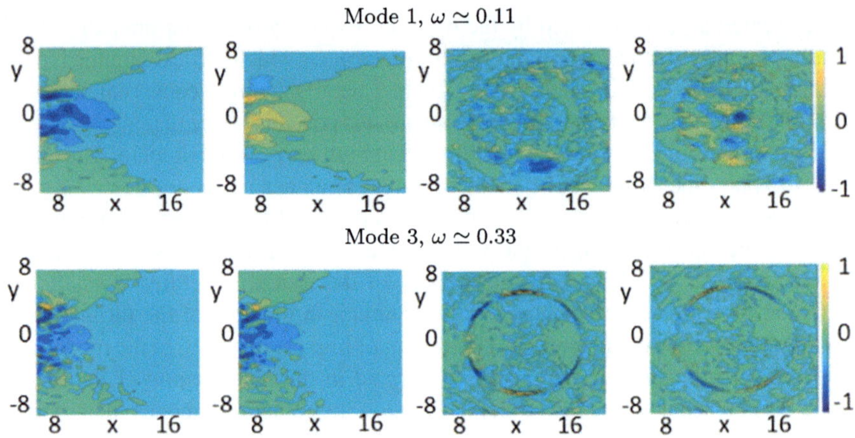

Fig. 3 DMD modes in the wake of a cross-flow wind turbine. From left to right: real and imaginary part of the streamwise velocity in the wake and the turbine

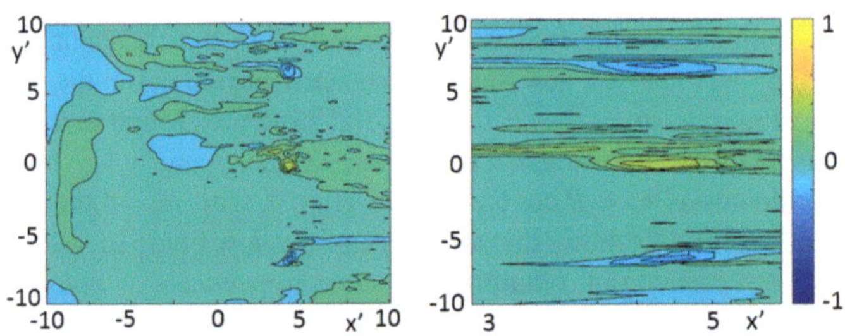

Fig. 4 Streamwise velocity in the wind turbine shown in Fig. 1 left in polar coordinates. Left: real size. Right: zoom in the area surrounded the turbine blades

Table 1 Frequencies of the DMD modes presented in Fig. 2

Wake		Turbine	
		Branch 1	Branch 2
ω_0	0.07	–	–
ω_1	0.11	–	0.10
ω_2	0.23	–	0.22
ω_3	0.33	0.33	–
ω_4	0.44	–	0.45
ω_5	0.54	–	0.54
ω_6	0.65	0.66	
ω_7	0.76	–	0.77
ω_8	–	–	0.90
ω_{10}	–	0.99	–

Regarding the spatio-temporal diagram of the wake of the wind turbine, shown in Fig. 1, one easily calculates that for a frequency $\omega_3 \simeq 0.33$ (turbine blades), the wavenumber of the leading spatio-temporal mode is $\kappa = \omega_3/c \simeq 0.5$ ($c \simeq 17$ previously approximated), in good agreement with the dominant wavenumber calculated in the turbine. This result suggest that this traveling wave, found in the wind turbine and the wake, is responsible for the asymmetry presented in the wake of the wind turbine.

4 Spatio-Temporal Koopman Decomposition

Spatio-temporal Koopman decomposition (STKD) [7] is a method that has been recently introduced to describe flow structures as a combination of traveling waves. The method is an extension of HODMD that applies such algorithm along the temporal and spatial components simultaneously with the aim of obtaining relations between spatial and temporal DMD modes. These new modes are called STKD modes q_{mn}, representing traveling waves, whose phase velocity is defined as $c = -\omega_n/\kappa_m$.

STKD decomposes spatio-temporal data $u(x, y, z, t)$ (where x, y, z and t are the streamwise, normal and spanwise components and t is the time) as a sum of spatio-temporal STKD modes in the following way:

$$u(x, y, z, t) \simeq \sum_{m,n=1}^{M,N} a_{mn} q_{mn}(y, z)\, e^{(\nu_m + i\kappa_m)x + (\delta_n + i\omega_n)t} \tag{2}$$

where ν_m and δ_n are the growth rates related to the spatial and temporal modes, respectively, and a_{mn} are the spatio-temporal amplitudes. Spatio-temporal diagrams $\omega - \kappa$, become determined by the value of the spatio-temporal amplitudes, determining the number of spatio-temporal STKD modes retained in the expansion. Using this method, it is possible to represent highly complex spatio-temporal structures as a group of traveling waves, helping in this way to understand complex physics.

The main algorithm considers two main steps, particularized for this problem they read as the following.

- As a first step, HOSVD is performed to a tensor U containing the original data in order to clean spatial redundancies or noise. HOSVD performs singular value decomposition (SVD) in each one of the spatial directions and combines the results. Hence, the original data tensor is decomposed in spatial modes in x, in y and in z and temporal modes in t. The number of retained SVD modes is established by a tolerance ε, set by the user, determining the root mean square (RMS) error of the approximated solution.
- As a second step, HODMD is applied to the temporal and spatial SVD modes in t and x, previously obtained. At this step, the main frequencies and wavenumbers are calculated. The number of retained modes is established by a second tolerance

ε_1, also set by the user. In addition, the parameter d for the HODMD analysis, named as d_s and d_t for the spatial and temporal analyses respectively, is also set by the user after some calibration based on the robustness of the results obtained.

5 STKD in Horizontal Axis Wind Turbines

Three-dimensional numerical simulations have been carried out with the aim of studying the wake of an horizontal axis wind turbine. The numerical code solves incompressible Navier–Stokes equations in non-linear form using spectral elements as spatial discretization. The codes uses cylindrical coordinate system, defined as (x, r, θ). The computational domain is a cylinder with dimensions $L_x = 63$ and $R = 35$. The $x - r$ plane is discretized using 4722 elements, each one discretized using Gauss-Labatto-Legendre basis functions with polynomial order 6. Finally, a Fourier decomposition is used along the azimuthal direction using 96 Fourier modes. The Reynolds number is set to Re $= 1000$.

The wind turbine is modeled using an actuator disk, where the thrust promoted by the turbine is rotationally symmetric and uniform over the disk. A uniform external forcing is added to the momentum equations of the fluid flow in the disk region. More details about these numerical simulations can be found in [18].

The wake of a wind turbine is mainly driven by the big bulk of fluid that travels downstream, following the wind direction, and that is rotating, due to the influence of the wind turbine axis rotation. These two main movements, longitudinal and rotational, make that the description and evolution of velocity fluctuations in the wake of a wind turbine are complex and, consequently, unpredictable and unknown. A good way to simplify reality is to study the flow evolution as a simple combination of traveling waves. To this purpose, it is necessary first, to study the spatio-temporal behaviour of the flow.

Figure 5 shows the spatio-temporal diagram of the streamwise velocity fluctuations downstream the wind turbine at a representative point of the computational domain. The figure shows two types of flow behaviour. In the near field,

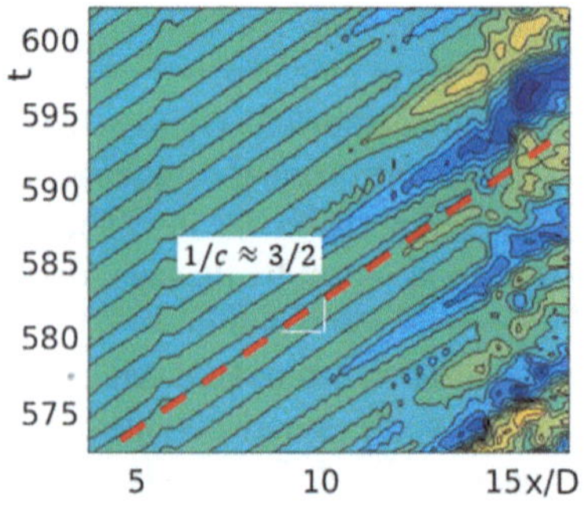

Fig. 5 Spatio-temporal diagram (time versus streamwise direction) along the streamwise direction of the streamwise velocity fluctuations at $r/D = 1$ and $\theta = \pi/4$

$1 < x/D < 13$, the figure suggests that there is a wave traveling along the streamwise direction, with slope $\Delta t/\Delta x \simeq 3/2$. In the far field, $x/D > 13$, it is possible to identify a change in the flow behaviours, probably related to the flow transition, although the presence of the traveling wave depicted in the near field is noticed.

STKD analysis has been performed to the three components of the velocity fluctuations in a smaller region of the computational domain, where the value of the main fluctuations starts growing ($3.5 < x < 18$). The analysis performed is three-dimensional, with the aim of finding the two-dimensional modes related to the main traveling waves that move along the streamwise direction, with phase velocity $c = -\omega_n/\kappa_m$.

A set of 100 snapshots, equispaced in time $\Delta t = 0.2$, has been used in the temporal STKD analysis, while a set of 240 points, equispaced in space along the streamwise direction $\Delta x = 5.8543 \cdot 10^{-2}$, is used for the spatial analysis. The parameters used for this analysis are $d_t = 20$ and $d_s = 40$, for the spatial and temporal analysis, respectively, and the tolerances are set to $\varepsilon = \varepsilon_1 = 0.01$ in a first analysis and $\varepsilon = 0.01$ and $\varepsilon_1 = 5 \cdot 10^{-3}$ for a second analysis. In the first case, STKD retains 7 spatio-temporal modes (plus their complex conjugate). As seen in Fig. 6 top left, these modes correspond to the main traveling wave presented in Fig. 5 and some other smaller amplitude traveling waves moving in the same direction but with

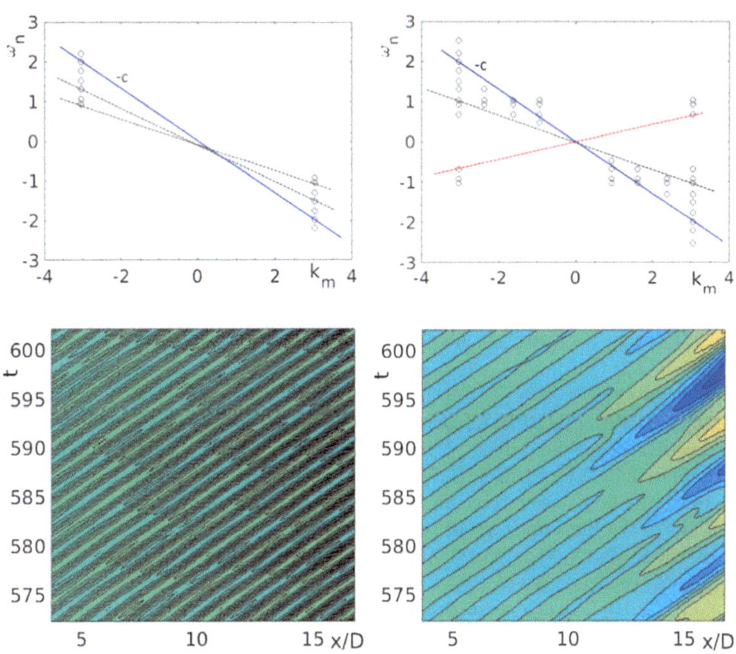

Fig. 6 Top: Frequencies versus wavenumber calculated in STKD using the indexes $d_t = 20$ and $d_s = 40$. Tolerances $\varepsilon = \varepsilon_1 = 0.1$ (left) and $\varepsilon = 0.01, \varepsilon_1 = 5 \cdot 10^{-3}$ (right). Bottom: reconstruction using the STKD expansion (2) using the modes presented in the top part of this figure

different phase velocity. Figure 6 bottom left shows the reconstruction of the velocity fluctuations using the aforementioned modes. As seen, the method represents the main traveling wave, however the flow representing the transient behavior is not well described. In the second case, STKD retains 22 spatio-temporal modes (plus their complex conjugate), shown in Fig. 6 top right. Most of these modes provide a better description of the main traveling wave (smaller amplitude spatio-temporal modes moving at different phase velocities but similar flow direction). Three spatio-temporal modes represent a traveling wave moving opposite to the wind direction. Figure 6 bottom right shows the reconstruction of the velocity fluctuations in this second case, where it is possible to distinguish the traveling wave and the transient flow, meaning that the traveling wave moving opposite to the wind direction is responsible for flow transition. More details about these results can be found in [4].

6 Conclusions

This Chapter presents a review on the applications of two methods for the analysis of flow structures in wind turbines. These methods, called HODMD and STKD, are able to simplify highly complex flows as a group of temporal modes oscillating in time and traveling waves. HODMD and STKD have been applied to detect the spatio-temporal patterns in a cross-flow and horizontal wind turbine, respectively. In both cases, the main results show that a traveling wave is responsible for triggering the transition from laminar to turbulent flows. With this review, the authors intend to show the community some alternative types of analyses for the detection of flow patterns and their application to wind turbines.

References

1. Ferrer E, Le Clainche S (2015) Flow scales in cross-flow turbines. In: Ferrer E, Montlaur A (eds) CFD for wind and tidal offshore turbines, Chap1. Springer Tracts in Mechanical Engineering, pp 1–11
2. Le Clainche S, Ferrer E (2018) A reduced order model to predict transient flows around straight bladed vertical axis wind turbines. Energies 11(3):566
3. Ramos G, Beltrán V, Le Clainche S, Ferrer E, Vega JM (2018) Flow structures in the turbulent wake of a cross-flow wind turbine. In: AIAA SciTech forum, wind energy symposium, AIAA paper-2018-0253. https://doi.org/10.2514/6.2018-0253
4. Le Clainche S, Mao X, Vega JM (2018) Traveling waves describing the wake of a wind turbine. Wind Energy (under review)
5. Schmid PJ (2010) Dynamic mode decomposition of numerical and experimental data. J Fluid Mech 656:5–28
6. Le Clainche S, Vega JM (2017) Higher order dynamic mode decomposition. SIAM J Appl Dyn Syst 16(2):882–925
7. Le Clainche S, Vega JM (2017) Spatio-temporal Koopman decomposition. Submitted to J Nonlinear Sci

8. Le Clainche S, Vega JM, Soria J (2017) Higher order dynamic mode decomposition for noisy experimental data: the flow structure of a zero-net-mass-flux jet. Exp Therm Fluid Sci 88:336–353
9. Le Clainche S, Vega JM (2017) Higher order dynamic mode decomposition to identify flow patterns and extrapolation properties. Phys Fluids 29(8):084102
10. Le Clainche S, Moreno R, Taylor P, Vega JM (2018) New robust method to study flight flutter testing. J Aircraft (in press)
11. Sirovich L (1987) Turbulence and the dynamic of coherent structures, parts I–III. Q Appl Math 45(3):561
12. Tucker LR (1996) Some mathematical notes on three-mode factor analysis. Psikometrica 31:279–311
13. Takens F (1981) Detecting strange attractors in turbulence. In: Rand DA, Young L-S (eds) Lecture notes in mathematics. Springer, Berlin, pp 366–381
14. Ferrer E (2017) An interior penalty stabilised incompressible discontinuous Galerkin-Fourier solver for implicit large Eddy simulations. J Comput Phys 348:754–775
15. Ferrer E, Willden RHJ (2010) A high order discontinuous Galerkin finite element solver for the incompressible Navier-Stokes equations. Comput Fluids 46(1):224–230
16. Ferrer E, Willden RHJ (2012) A high order discontinuous Galerkin-Fourier incompressible 3D Navier-Stokes solver with rotating sliding meshes. J Comput Phys 231(21):7037–7056
17. Le Clainche S, Sastre F, Vega JM, Velazquez A (2017) Higher order dynamic mode decomposition applied to study flow structures in noisy PIV experimental data AIAA 2017-3304. In: Proceedings of 47th AIAA fluid dynamics conference. Denver, CO, USA, 5–9 June
18. Mao X, Sorensen J (2018) Far-wake meandering induced by atmospheric eddies in flow past a wind turbine. J Fluid Mech 846:190–209

Towards Numerical Simulation of Offshore Wind Turbines Using Anisotropic Mesh Adaptation

L. Douteau, L. Silva, H. Digonnet, T. Coupez, D. Le Touzé and J.-C. Gilloteaux

Abstract In the context of reducing the cost of floating wind energy, predicting precisely the loads applied on structures and their response is essential. As the simulation of floating wind turbines requires the representation of both complex geometries and phenomena, several techniques have been developed. The wake generated by the aerodynamic loads experienced and the tower can be modeled using methodologies inherited from onshore wind simulation, and coupled with a hydrodynamic codes that were most of the time developed for the oil and gas industry. This work proposes a methodology for the simulation of a single or several turbines with an exact representation of the geometries involved, targeting an accurate evaluation of loads. The software library used is ICI-tech, developed at the High Performance Computing Institute (ICI) of Centrale Nantes. A single computational mesh is used, where every phase is defined through level-set functions. The Navier–Stokes (NS) equations are solved in the Variational MultiScale (VMS) formalism using finite element discretization and a monolithic approach. A coupling with an automatic and anisotropic adaptation procedure guarantees the good representation of the geometries immersed. The adaptation allows the simulation of phenomena

L. Douteau (✉) · L. Silva · H. Digonnet · T. Coupez
High Performance Computing Institute, Ecole Centrale de Nantes,
1 rue de la Noë, 44300 Nantes, France
e-mail: louis.douteau@ec-nantes.fr

L. Silva
e-mail: luisa.rocha-da-silva@ec-nantes.fr

H. Digonnet
e-mail: hugues.digonnet@ec-nantes.fr

T. Coupez
e-mail: thierry.coupez@ec-nantes.fr

D. Le Touzé · J.-C. Gilloteaux
Research Laboratory in Hydrodynamics, Energetics and Atmospheric Environment,
Ecole Centrale de Nantes, 1 rue de la Noë, 44300 Nantes, France
e-mail: david.letouze@ec-nantes.fr

J.-C. Gilloteaux
e-mail: jean-christophe.gilloteaux@ec-nantes.fr

E. Ferrer and A. Montlaur (eds.), *Recent Advances in CFD for Wind and Tidal Offshore Turbines*, Springer Tracts in Mechanical Engineering,
https://doi.org/10.1007/978-3-030-11887-7_9

with very different orders of magnitude, e.g. aerodynamics around blades and waves propagation. The reduction of the number of points in the mesh and the massive parallelization of the code are also necessary for wind turbine simulation.

1 Introduction

Fast and accurate simulation is crucial for the development of floating wind energy. Simulating floating wind turbines under various aerodynamic and hydrodynamic cases will enable better dimensioning of the structures, which is crucial for industrial deployment. Motivated by the development of onshore wind energy, many authors studied the aerodynamic behavior of turbines, e.g. [1], often with rotor models reducing the computational effort. But those models struggle to keep their reliability in a floating context. Depending on the wave conditions, wave-induced motions may induce significant effects on the energy production. Hydrodynamic effects hold a huge influence, and [2] highlighted their impacts on energy production. However, dealing with a thin representation of a wind turbine under aerodynamic and hydrodynamic loads requires high efforts in both development and computations. Consequently, only few authors have shown interest in coupled simulations, e.g. [3, 4] or [5]. In a majority of cases, only a component was studied at a time, as in [6] or [7].

The full resolution of the geometries, boundary layers and hydrodynamic effects for floating wind turbines is a real challenge. Events with very different orders of magnitude for both characteristic time and length are observed in a large computational domain, and can be handled using several overlaid meshes with different levels of refinement, e.g. in [4]. This work proposes an alternative methodology based on an unique computational mesh. Both wind turbines and the air/water interface are immersed and defined using level-set function and mixing laws. The Navier–Stokes (NS) equations are solved using stabilized finite elements within a Variational-MultiScale (VMS) formulation, with a monolithic approach. The computational mesh is adapted anisotropically and automatically during the simulation.

The methodology used in this work is detailed in Sect. 2. The first results obtained are presented in Sect. 3.

2 Methodology

2.1 Mesh Immersion Procedure

The mesh immersion technique uses a level-set approach presented in [8] to represent immersed bodies. A signed-distance α is measured at each point of the computational mesh, which allows to build the smoothed Heaviside function H_ε shown in Eq. (1). A width parameter ε is introduced.

Fig. 1 Left: signed-distance α. Center: smoothed Heaviside H_ε. Right: adapted mesh

$$\begin{aligned} H_\varepsilon(\alpha) &= \tfrac{1}{2}\left(1 + \tfrac{u_\varepsilon(\alpha)}{\varepsilon}\right), \\ u_\varepsilon(\alpha) &= \varepsilon \tanh\left(\tfrac{\alpha}{\varepsilon}\right). \end{aligned} \tag{1}$$

This immersion technique introduces a transition area around the frontier of the immersed object. The results are strongly dependent on the mesh, and coupling this approach with an automatic mesh generation procedure handles this dependency. An example of reconstruction is proposed in Fig. 1. The results obtained with H_ε provide an interior, depicted in black. The quality of the reconstruction is due to the adapted mesh.

The computational complexity brought by this methodology becomes considerable when a M-elements mesh is immersed in a N-points computational mesh. The complexity of the level-set approach is $N \times M$, but could be reduced through the construction of an octree, within a minimal theoretical complexity of $N \times \log(M)$.

2.2 *Anisotropic Mesh Adaptation*

The mesh adaptation procedure is essential for the reconstruction as presented in Sect. 2.1, but is also a self-standing software unit critical for ICI-tech. The computational mesh is automatically generated from a coarse initial mesh and tends to gather points in interest zones. The theory proposed by [9] is based on a principle of equidistribution of the error in all the computational domain. An a posteriori error estimator is evaluated along the edges of the mesh, important variations highlighting needs for adaptation. Scaling factors are defined along each dimension, which allows to anisotropically deform the mesh. This procedure is repeated iteratively, until an adapted mesh is obtained.

The immersion can be dependent on the elements immersed into the computational domain. However, this criteria prevents a good resolution of fluid flows, as large cells introduce numerical dissipation and miss sub-scale flows. The error estimator is modified to take into account both the immersed bodies and the velocity.

2.3 Mixing Laws

The resolution of fluid flows in ICI-tech is achieved from the incompressible NS equations. Viscosity and density are interpolated at each point of the computational mesh, thanks to mixing laws. A linear law is presented in Eq. (2) for the immersion of an object of viscosity η_i in a domain of viscosity η. If several elements are immersed, the mixing laws are applied successively. Note that when large disparities are encountered, other laws may be more appropriated, e.g. logarithmic ones.

$$\eta_p(\alpha) = \eta_i H_\varepsilon(\alpha) + \eta(1 - H_\varepsilon(\alpha)). \tag{2}$$

2.4 Resolution of the Navier–Stokes Equations

ICI-tech proposes a monolithic approach based on stabilized finite elements where the incompressible NS problem, presented in a variational form Eq. (3) with $\mathbf{v}$ velocity and p pressure, is solved in a VMS formulation.

$$\begin{aligned}&\forall(\mathbf{w}, q) \in V_0 \times Q,\\ &\begin{cases}\rho(\partial_t \mathbf{v}, \mathbf{w}) + \rho(\mathbf{v} \cdot \nabla\mathbf{v}, \mathbf{w}) + 2\nu\boldsymbol{\varepsilon}\mathbf{v} : \boldsymbol{\varepsilon}\mathbf{w} - (p, \nabla \cdot \mathbf{w}) = (\mathbf{f}, \mathbf{w})\\ (\nabla \cdot \mathbf{v}, q) = 0\end{cases}\end{aligned} \tag{3}$$

The VMS paradigm operates as an implicit-LES (Large Eddy Simulation). Velocities are split between the coarse scales, which are solved, and sub-grid scales, which are modeled. This writes $\mathbf{v} = \mathbf{v}_h + \mathbf{v}'$, with $\mathbf{v}_h$ the coarse scale velocity and $\mathbf{v}'$ the sub-grid scale one. In LES, a viscous term is incorporated in the NS equations to compensate for the effect of the subgrid-scale turbulence. The VMS reformulates the NS problem to express implicitly the subgrid-scale velocities in the formulation, and to mathematically evaluate their influences. The VMS formulation of the NS problem features stabilization terms (τ_K, τ_C) and residuals ($\mathscr{R}_\mathbf{M}$, $\mathscr{R}_C$) coming from this evaluation of subgrid-scales. More details can be found in [9].

$$\begin{aligned}&\forall(\mathbf{w}_h, q_h) \in V_{0,h} \times Q_h,\\ &\begin{cases}\rho(\partial_t \mathbf{v}_h, \mathbf{w}_h) + (\rho\mathbf{v}_h \cdot \nabla\mathbf{v}_h, \mathbf{w}_h) - \sum\limits_{K\in\mathscr{T}_h} (\tau_K \mathscr{R}_\mathbf{M}, \rho\mathbf{v}_h \nabla\mathbf{w}_h)_K\\ +2\mu\boldsymbol{\varepsilon}(\mathbf{v}_h) : \boldsymbol{\varepsilon}(\mathbf{w}_h) - (p_h, \nabla \cdot \mathbf{w}_h) + \sum\limits_{K\in\mathscr{T}_h} (\tau_C \mathscr{R}_C, \nabla \cdot \mathbf{w}_h)_K = (\mathbf{f}, \mathbf{w}_h)\\ (\nabla \cdot \mathbf{v}_h, q_h) - \sum\limits_{K\in\mathscr{T}_h} (\tau_K \mathscr{R}_\mathbf{M}, \nabla q_h)_K = 0\end{cases}\end{aligned} \tag{4}$$

This formulation is then discretized in time using an Euler implicit scheme. The non-linear terms which appeared are linearized thanks to a Newton approach, and the final linear system is solved using PETSc.

2.5 Simulation Procedure

The simulations with ICI-tech rely on mesh adaptation, phase reconstruction and resolution of the NS equations, detailed in the previous subsections. The procedure starts with a mesh adapted on immersed objects, obtained after immersion iterations. The reconstruction of the phases is performed with this mesh, and mixing laws are applied. The NS problem is then solved with a given time step in velocity and pressure. The mesh is adapted from weighted velocities and level-set results. This procedure is repeated until full simulation is performed.

Computational costs are optimized, through a limitation of the complexity of the algorithms, e.g. with the octree discussed in Sect. 2.1, along with massive parallelization. Particular attention has been placed on partitions during parallel meshing and NS resolution, details can be found in [10].

2.6 Computing the Force Applied on an Immersed Element

The computation of the force applied on an object Ω is not trivial in the context of mesh immersion, as the adapted mesh obtained is not body fitted. However the concentration of points around the interfaces enables an accurate evaluation of the loads. The forces are computed a posteriori from an integration of the local constraint $\mathbf{T}_{local}$ written in Eq. (5) with η the viscosity and $\mathbb{I}$ the identity matrix, using an approach similar to the one proposed in [11].

$$\mathbf{T}_{local} = \sigma \cdot \mathbf{n}, \text{ with } \begin{cases} \sigma = \eta(\nabla \mathbf{v} + \nabla \mathbf{v}^T) - p\mathbb{I} \\ \mathbf{n} = -\frac{1}{\|\nabla u_\varepsilon\|} \nabla u_\varepsilon \end{cases} \tag{5}$$

$$\mathbf{F} = \int_{\partial\Omega} \mathbf{T}_{local} dS \tag{6}$$

The expression of $\mathbf{F}$ is reformulated to turn the integral over $\partial\Omega$ into an integral over Ω. The force computed over the transition zone, F_ε, is presented in Eq. (7).

$$\mathbf{F}_\varepsilon = \int_{\Omega} \delta_\varepsilon \mathbf{T}_{local} dV, \text{ with } \delta_\varepsilon = \frac{1}{2\varepsilon}\left(1 - \left(\frac{u_\varepsilon}{\varepsilon}\right)^2\right) \tag{7}$$

This formulation is inherited from the smoothed Heaviside function. The span of the integration zone is defined by ε, and the Dirac function introduced δ_ε is derived from u_ε. The convergence of $\mathbf{F}_\varepsilon$ towards $\mathbf{F}$ when the width of the transition area tends towards zero has been discussed in [11]. Consequently, the quality of the force computed is determined by the accuracy achieved within the immersion process.

3 Application to Floating Wind Turbines

3.1 Floating Wind Turbines Reconstruction

The reconstruction of floating wind turbines has been studied in order to evaluate the number of nodes required for an accurate representation of the geometries. The surface mesh of a prototype studied in wind flume by [12], composed of approximatively 75 K facets and depicted in Fig. 2, is immersed in a computational domain of 120 K points. The order of precision of the immersion is of the decimeter for a full-scale wind turbine. This precision, appropriated for visualization purposes, can be improved for accurate flow simulations. A slice of the mesh highlights the presence of coarse cells far from the turbines, which will be refined during the simulation to capture the wake.

As the reconstruction needs to be performed at each time step, optimization is required. The computational savings provided by the octree have been quantified. Table 1 presents the reconstruction of a single wind turbine in an adapted mesh. The construction of the octree was performed with a limit of 200 elements per subdomain and a maximal depth of 12, and the number of distances to evaluate at each time step are measured. Computational cost has been largely reduced thanks to the addition of this octree, which confirms the interest of this type of structure for the distance computation algorithm, and by extension for the mesh immersion procedure.

The scalability of the whole reconstruction procedure has also been tested. Floating wind turbines described by the mesh depicted in Fig. 2 were reconstructed with a constant precision and different numbers of processors. 50 increments of adaptation from a coarse initial mesh were perform, with constant octree parameters. Table 2 presents the results from a weak speed-up study performed with a constant number of 6 wind turbines per processor. A hard speed-up study was also performed, with the reconstruction of 10 wind turbines with a mesh of 250 K points and

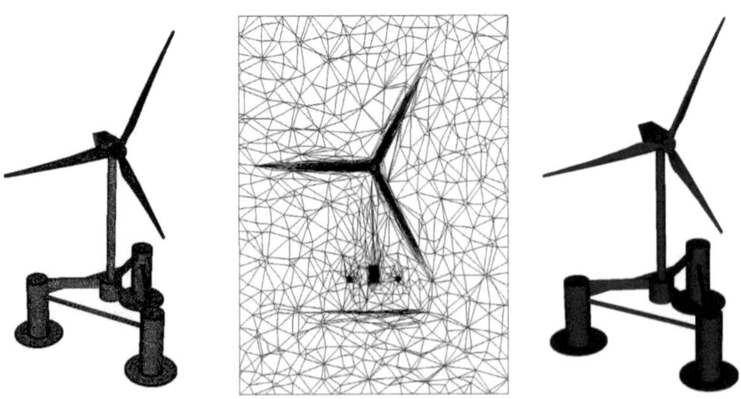

Fig. 2 Left: immersed mesh. Center: slice of adapted mesh. Right: reconstructed turbine

Table 1 Influence of the octree implementation on wind turbine immersion

	Distances evaluated	Time for distance evaluation	Time per distance (s)
No Octree	~8693 M (100%)	9 min, 51 s (100%)	$6.8 \cdot 10^{-8}$s
Octree	~7 M (0.081%)	2.5 s (0.4%)	$3.5 \cdot 10^{-7}$s

Table 2 Weak speed-up study with 6 WTs per processor

Case	Nodes/core (K)	Adaptation (s)	Immersion (s)	I/O (s)
1 WT	25	6.2 K	62	7.1
10 WTs	24.9	8.2 K (×1.3)	119 (×1.9)	20.1 (×2.8)
100 WTs	26.5	9.8 K (×1.6)	667 (×10.8)	140 (×19.7)

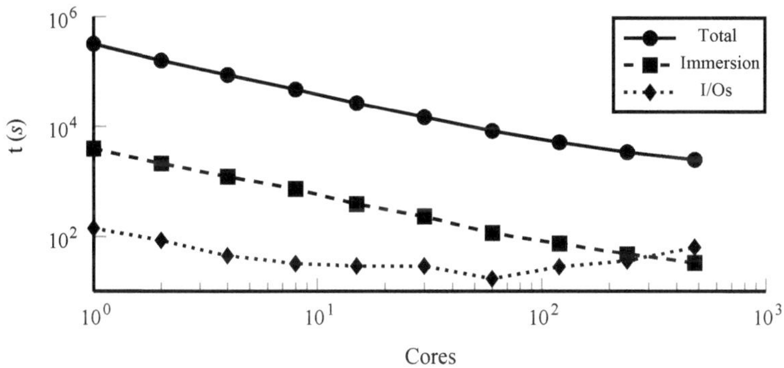

Fig. 3 Time required to reconstruct 10 wind turbines with different number of processors

different numbers of processors. The results are presented in Fig. 3. The adaptation and immersion proved to be very scalable, while the I/Os tended to limit the efficiency of the immersion for low load per core.

3.2 *First Validation Towards High Reynolds Number Single-Phase Flows*

ICI-tech is currently used for applications in material forming, on low-Reynolds number flows. In the context of wind turbines, the combination of wind speed and blade rotation speed generates flows at the tip of blades with Reynolds numbers of several millions. The validation of the flow solver is currently conducted to extend its scope to wind turbine applications. To that extent, drag and lift coefficients C_p and C_l are computed, from Eq. (8), with A the cross sectional area and $\mathbf{F}$ the force computed with the method presented in Sect. 2.6.

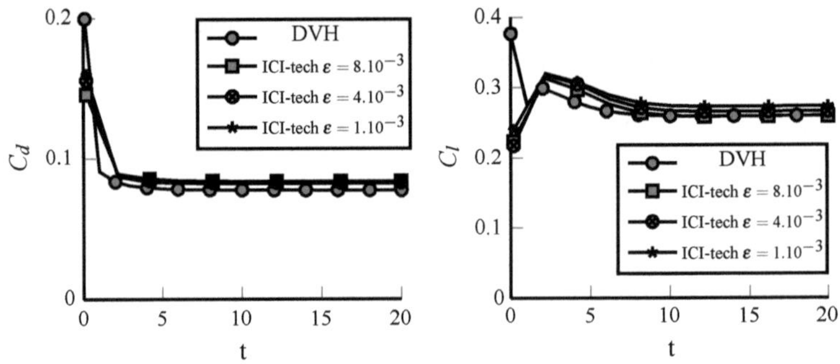

Fig. 4 Flow passing a NACA-0008 at $Re = 2000$. Left: drag coefficient. Right: lift coefficient

$$C_d = \frac{2}{\rho U^2 A}\mathbf{F} \cdot \mathbf{x}\ ;\ C_l = \frac{2}{\rho U^2 A}\mathbf{F} \cdot \mathbf{y} \tag{8}$$

The verification started on 2D test cases, where moderate Reynolds flows passing NACA profiles were studied. The results obtained with ICI-tech were compared against data from [13]. Validation was performed on test cases for which the Reynolds number is of the order of a thousand, and the influence of different parameters used in computations are highlighted. A transition was performed towards 3D, by extending the referred test case. Flow passing a NACA-0008 with an orientation of 4° immersed in a constant uniform flow of $Re = 2000$ was studied. The results obtained are compared in Fig. 4 for different ε and a constant number of nodes in the computational mesh.

At the steady state, errors remain for both drag and lift or pressure coefficients. Decreasing ε has little influence on the precision obtained with the drag coefficient, while the convergence in ε for lift does not tend towards DVH results. Further research is currently conducted in order to improve the results presented, in particular a solver reimplementation focusing on high-Reynolds flows is planned.

In the following, unsteady test cases at moderate Reynolds numbers are studied. This is required in order to evaluate the capacities of the solver to handle vortex shedding, critical in wind turbine simulations. However, those $Re = O(1000)$ flows are not representative of the phenomena occurring around wind turbine blades. The validation will then continue on higher Re, e.g. with a comparison with experimental results from [14]. The validation on single-phase flows will be ended with simulations on blades and wind turbine rotor in movement.

3.3 Wave Generation and Free-Surface Tracking

The previous subsection presented the steps required to obtain a validated monophasic solver, which paves the way for accurate simulations on wind turbines. However,

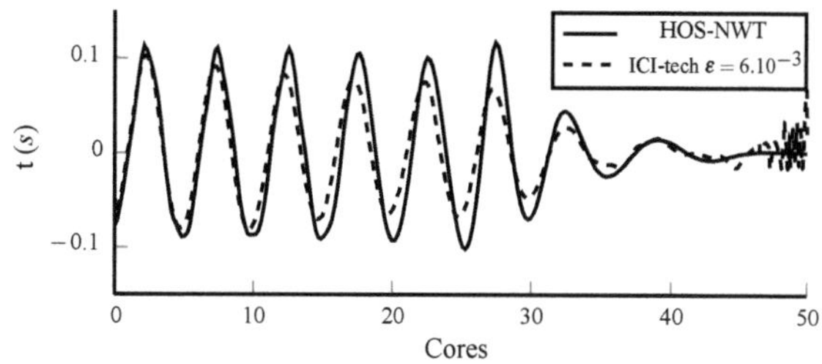

Fig. 5 Time required to reconstruct 10 wind turbines with different number of processors

the floating characteristic imposes triphasic simulations, with an the presence of a free-surface between air and water. The tracking of this interface is realized with the techniques presented in Sect. 2. Each phase is represented using level-set functions, and the NS equations are solved in the domain. The convection of the level-set representing the free-surface is realized by means of the convected level-set method detailed in [15].

A numerical tank was implemented, with a piston-type wave maker whose movements are determined using HOS-NWT [16]. The verification consisted on the tracking of the free-surface generated. A wave field of period 0.55s and amplitude 0.1m propagated from the left of the computational domain is drawn in Fig. 5. A ε of 6.10^{-3} and 40 K points are used for the computational mesh. The free-surface obtained with ICI-tech is drawn in blue, and the target wave field input in red. Simulations were realized with viscosities a hundred-times bigger than the reality for each phase. This artefact enabled to better stabilize the computations, but will have to be suppressed in order to get high precision results.

If the fidelity of the first wave generated is interesting, the ICI-tech free-surface results progressively feature more damping and a very important noise at the right end of the domain. The damping can be partially explained by the higher viscosities chosen for the fluids, but the influence of numerical diffusion will have to be overlooked. The noise observed will be removed by the addition of a numerical absorption area at the right end of the domain. A transition towards a 3D numerical wave tank is being performed, with the final objective to be able to generate complex wave fields.

4 Conclusion and Outlook

ICI-tech shows a potential to accurately simulate floating wind turbines, even if further developments are still needed. The coupling between the computational mesh and the immersion procedure provides a lot of flexibility to the simulations, and limits the size of the computational meshes. The costs commonly required for full-scale LES simulations can be reduced, while a similar precision on the results can be obtained thanks to the VMS solver.

The reconstruction procedure has been optimized, and a satisfying scalability has been obtained. Further developments are concentrated on the solver. Validation is ongoing, both for monophasic and multiphasic. A transition towards high-Reynolds flows and wind turbine blades is planned, and free-surface simulations will be extended to regular and irregular wave fields. Fluid-structure interactions will then be required to present a wind turbine demonstrator.

Acknowledgements This work is funded by the WEAMEC (West Atlantic Marine Energy Community), and was performed by using HPC resources of the Centrale Nantes Supercomputing Centre on the cluster Liger and supported by a grant from the Institut de Calcul Intensif (ICI) under the project ID E1611150/2016.

References

1. Hansen MOL et al (2006) State of the art in wind turbine aerodynamics and aeroelasticity. Prog Aerosp Sci 42:285–330
2. Sebastian T, Lackner M (2011) Offshore floating wind turbines - an aerodynamic perspective. In: 49th AIAA aerospace sciences meeting including the new horizons forum and aerospace exposition, vol 720
3. Leble V, Barakos G (2016) Demonstration of a coupled floating offshore wind turbine analysis with high-fidelity methods. J Fluids Struct 62:272–293
4. Quallen S, Xing T (2016) CFD simulation of a floating offshore wind turbine system using a variable-speed generator-torque controller. Renew Energy 97:230–242
5. Yan J et al (2016) Computational free-surface fluid-structure interaction with application to floating offshore wind turbines. Comput Fluids 141:155–174
6. Tran T-T, Kim D-H (2015) The platform pitching motion of floating offshore wind turbine: a preliminary unsteady aerodynamic analysis. JWEIA 142:65–81
7. Wu CHK, Nguyen VT (2017) Aerodynamic simulations of offshore floating wind turbine in platform induced pitching motion. Wind Energy 20:835–858
8. Coupez T et al (2015) Implicit boundary and adaptive anisotropic meshing. In: New challenges in grid generation and adaptivity for scientific computing, pp 1–18
9. Coupez T, Hachem E (2013) Solution of high-Re. incompressible flow with stabilized finite element and adaptive anisotropic meshing. CMAME 267:65–85
10. Digonnet H et al (2017) Massively parallel anisotropic mesh adaptation. IJHPCA 33(1):3–24
11. Brackbill J et al (1992) A continuum method for modeling surface tension. J Comput Phys 100:335–354
12. Lacaze, J.-B. et al. Small scale tests of floating wind turbines in the wind and wave flume of Luminy. *14th journées d'hydrodynamique*, Val de Reuil, France, 18–20 Nov 2014
13. Rossi E et al (2016) Simulating 2D viscous flow around geometries with vertices through the Diffused Vortex Hydrodynamics method. CMAME 302:147–169
14. Bak C et al (2000) Wind tunnel tests of the NACA 63-415 and a modified NACA 63-415 airfoil. Forskningscenter Risoe, Risoe-R, Denmark, No. 1193
15. Ville L et al (2011) Convected level set method for the numerical simulation of fluid buckling. IJNMF 66:324–344
16. Ducrozet G et al (2012) A modified high-order spectral method for wavemaker modeling in a numerical wave tank. EJMBF 34:19–34

Numerical Modelling of a Savonius Wind Turbine Using the URANS Turbulence Modelling Approach

Tomasz Krysinski, Zbigniew Bulinski and Andrzej J. Nowak

Abstract This work presents a three-dimensional investigation of the performance prediction of the operation of the vertical axis wind turbine. The analysis was carried out for the micro-turbine equipped with the Savonius rotor. The applied methodology was based on the Computational Fluid Dynamics (CFD) and used the Finite Volume method to solve the unsteady Reynolds Averaged Navier–Stokes equations. We concentrated our investigations on the influence of the turbulence modelling methodology on the simulation results. The analysis considers most of the URANS turbulence models, starting with the commonly used two equation models like $k-\epsilon$ or $k-\omega$ SST to the Reynolds Stress Models with quadratic pressure strain modelling. The results show the influence of turbulence models on the results of the predicted flow field and wind turbine performance.

1 Introduction

Currently the wind energy production sector is dominated by the high power wind turbines with the horizontal rotation axis of the turbine rotor - HAWTs. Such units offer not only higher power but also higher efficiency in comparison to the units with the vertically oriented rotor - VAWTs [1–3]. Pope showed using Computational Fluid Dynamics tools that the common constructions of small HAWTs are approximately twice as efficient as a small drag-based VAWTs constructions [4]. However, some authors suggest that recently existing proportions between the horizontal and vertical units will undergo a significant change in the foreseeable future [5, 6]. Islam

T. Krysinski (✉) · Z. Bulinski · A. J. Nowak
Institute of Thermal Technology, Silesian University of Technology, Gliwice, Poland
e-mail: tomasz.krysinski@polsl.pl

Z. Bulinski
e-mail: zbigniew.bulinski@polsl.pl

A. J. Nowak
e-mail: andrzej.j.nowak@polsl.pl

E. Ferrer and A. Montlaur (eds.), *Recent Advances in CFD for Wind and Tidal Offshore Turbines*, Springer Tracts in Mechanical Engineering,
https://doi.org/10.1007/978-3-030-11887-7_10

et al predict that it is possible that in the next two or tree decades the vertical axis constructions will dominate the wind energy sector [7].

The fluid flow around the operating wind turbine, irrespective of the particular design type, is unsteady, complex and highly turbulent. The correct prediction of the flow structures occurring in the proximity of the rotating wind turbine is crucial for the accurate prediction of the turbine performance and operation, including forces acting on the turbine and noise generated by the turbine. Therefore, reliable methods for studying the behaviour of the wind turbines are something that is solely needed to solve engineering and safety issues arising at the designing and construction of the wind turbine. As an effect of the rapid development in the high performance computing sector, the research methods based on the computational fluid dynamics (CFD) approach are now more popular than ever. Therefore, natural direction in the engineering practise is to apply advanced computer tools in the process of designing and construction of the wind turbine. But, once the simulation tool is to be applied to design any machinery, the issue of the reliability of the simulation results arises. Therefore, there is a great need to carry out credibility analysis of the numerical models of the wind turbine operation. In this work we analyse influence of the turbulence model on the prediction of the mathematical model describing operation of the Savonius type wind turbine. This type of turbine is commonly, considered as a drag-based wind turbine, because the main driving force is the drag force. The flow around this type of turbines is isothermal and incompressible, however its unsteadiness and highly turbulent character make it difficult to model.

There are three main modelling approaches to describe turbulent flows [8]:

- Unsteady Reynolds Averaged Navier–Stokes - URANS.
- Large Eddy Simulations - LES.
- Direct Numerical Simulations - DNS.

It would be extremely difficult to resolve the whole flow field around operating wind turbine with the use of DNS or LES techniques in a reasonable time with currently available computational resources. Therefore, only the RANS methodology is available for practical application in wind turbine design. Currently, the most popular turbulent models applied to describe flow around wind turbine are the URANS models.

In this chapter authors present comparisons of the accuracy of the various URANS turbulence models in the 3D modelling of the Savonius type wind turbine. The primary concerns in the studies were the turbine performance prediction, as well as the flow structures propagation.

2 Mathematical Model of VAWT

The proper mathematical model of the operating VAWT should include all the necessary physics of the dynamic interactions between the turbine rotor and the air inflow. Due to low values of the Mach number, which is far below conventional limiting value of 0.3, the air flowing around a typical Savonius turbine can be treated as an incompressible fluid. The presence of a wide range of temporal and spatial turbulence scales introduces significant difficulties in developing the credible numerical model. Hence, crucial is the choice of the appropriate turbulence model as well as the proper size of the time step and numerical grid which should be adjusted to the turbulence model settings.

By its principle the RANS approach to turbulence modelling assumes that instantaneous values of flow variables are sums of their averaged values and fluctuations. For steady state problems, the averaging process may be considered as a simple temporal averaging. But this approach is difficult to apply for unsteady problems because it is difficult to choose the averaging period length which would provide good statistics on one hand and it would not average out unsteadiness of the flow on the other. Therefore, in the case of the transient flows, the flow averaging process is carried out by the use of the ensemble average operator, this is commonly known as the URANS methodology.

The ensemble average operator can be interpreted as an expected value of N independent realisations of exactly the same flow which would have exactly the same unambiguity conditions: domain geometry, initial and boundary conditions and so on. Applying the ensemble average operator to a single realization of a flow variable (assuming a flow variable can be treated as a random variable) results in the following decomposition [8]:

$$\phi(x,t) = \overline{\phi}(x,t) + \phi'(x,t), \tag{1}$$

where $\overline{\phi}$ is considered as an average of an arbitrary flow variable and ϕ' as a fluctuation. Therefore, the flow describing equations in the URANS framework can be written as [8, 9]:

$$\nabla \cdot \overline{\mathbf{u}} = 0, \tag{2}$$

$$\rho \frac{\partial \overline{\mathbf{u}}}{\partial t} + \rho \nabla \cdot (\overline{\mathbf{u}}\overline{\mathbf{u}}) = -\nabla \overline{p} + \mu \nabla^2 \overline{\mathbf{u}} + \nabla \cdot \tau^R, \tag{3}$$

where ρ stands for the fluid density, $\mathbf{u}$ is the velocity vector, p denotes pressure, t stands for time and τ^R denotes Reynolds stress tensor. This tensor is defined as an average dyadic product of velocity fluctuation vectors, mathematically it can be interpreted as a covariance of the velocity oscillations. While, from the physical point of view it works as an additional stress due to the turbulent mixing of the fluid. Despite the fact that the Reynolds stress tensor is symmetric in its nature, it introduces six additional unknowns besides the pressure and velocity vector components. This computation of the components of the stress tensor using information on the average velocity field is commonly called The Closure Problem and it can be resolved by the introduction of an additional assumptions and equations. These closing assumptions are known as a turbulence model.

One of the most commonly used approach addressing this problem is so called Boussinesq analogy, performance of this approach was tested for numerous engineering problems giving reasonable results, that is why it is widely implemented in the commercially available CFD packages. The Boussinesq analogy in its foundation assumes the linear dependence between the Reynolds stress components and the gradient of the average velocity vector [9]:

$$\tau_{ij}^{R} = -\rho\overline{u'_i u'_j} = \mu_t\left(\frac{\partial\overline{u_i}}{\partial x_j} + \frac{\partial\overline{u_j}}{\partial x_i}\right) - \frac{2}{3}\rho k\delta_{ij} \tag{4}$$

where μ_t is turbulent viscosity coefficient, δ_{ij} is the Kronecker delta and k stands for turbulence kinetic energy. As it can be easily noted, the proportionality coefficient between Reynolds stress tensor and the mean strain rate of the flow is defined as turbulent viscosity or eddy viscosity. This often results in situation when the turbulence models that utilise Boussinesq analogy are called eddy viscosity models. Approaching the closure problem with such an assumption has its benefits, like simplicity that also allows to develop computationally efficient and robust turbulence models. This feature has greatly contributed to nowadays popularity of this type of models. However, it is crucial to state that Boussinesq analogy is a sort of approximation and assumes an isotropic character of turbulence. This implies that turbulence models based on this hypothesis may incorrectly predict fluid flow behaviour with strong turbulence anisotropy. Despite the popularity of the eddy viscosity models among engineers and researchers, not all type of flows can be correctly predicted by this type of modelling approach. In some engineering problems the Boussinesq analogy turns out to be insufficient. Particularly problematic to be modelled are: flows with boundary layer separation, rotating flows or flows along curved wall (see Wilcox [10]). These flow phenomena are to be partially expected in the air flow around an operating wind turbine. Without making Boussinesq analogy assumption, the individual terms of the Reynolds stress tensor need to be calculated directly. It is possible by closing the set of equations for turbulence model with the introduction of the second order moment equations, namely the Reynolds stresses transport equations. Such an operation is called the second order closure. The turbulence models that utilise this approach are called Reynolds Stress Models (RSM). By the definition, the RSMs can be described as the most advanced, complex and complete URANS type models. The difference between all others URANS turbulence models underlay in the method of calculation of the turbulent viscosity.

As it was already mentioned, the flow around a vertical axis wind turbine is strongly turbulent; hence, a proper turbulence modelling is one of the most important issues. In the literature, most researchers concentrate on selected turbulence models without conducting comparison studies for a wider range of models [11–20]. In the presented work the most important variants of the URANS turbulence models were utilised, purposely encompassing a range from popular and simple two equation models to the most advanced and complex Reynolds Stress Models.

3 Turbulence Modelling

Despite the apparent simplicity due to incompressibility and isothermal character of the airflow around the considered wind turbine rotor; in fact, many turbulent and complex flow structures can be observed. Due to the rotational movement of the turbine rotor air masses are constantly swirled and mixed. This behaviour is prominent especially in a case of the modelled Savonius wind turbine. The coupling of the physical phenomena like boundary layer separation, vortex shedding or dynamic stall can be observed. The nature of all these phenomena feeds the turbulence in the airflow surrounding the rotor of the turbine. Therefore, proper modelling of this complex phenomena is a crucial aspect of modelling the operating wind turbine.

The first group of utilised models were the eddy viscosity models. Turbulence models belonging to that group can be found in nearly all of the commercially available CFD software. Relatively cheap, efficient and reasonably accurate for a number of engineering applications. The set of eddy viscosity turbulence models which performance was verified in this work are as follows:

- $k - \epsilon$ realizable
- standard $k - \omega$
- $k - \omega$ SST
- Shear Stress Transport.

The special care was devoted to the performance of the second-order closure models. In this group four the most popular variations of the RSMs were considered:

- Linear Stress Model
- Quadratic Stress Model
- Stress-ω Model
- Baseline Stress Model - BSL.

The most notable difference between the eddy viscosity models and the RSMs is the approach in modelling the elements of the Reynolds Turbulent Stresses tensor. The use of the RSMs require calculation of all components of the Reynolds stress tensor is carried out. For this reason, a transport equation is formulated for each of 6 independent Reynolds stresses:

$$\frac{\partial \left(R_{ij}\right)}{\partial t} + \nabla \left[R_{ij}\mathbf{u}\right] = P_{ij} + D_{ij} - E_{ij} + \Pi_{ij} + \Omega_{ij} \tag{5}$$

where R_{ij} stands for Kinematic Reynolds Stresses equal to the fluctuations covariance $\overline{u_i'u_j'}$, P_{ij} is the stress production term due to mean flow gradient, D_{ij} is the stress diffusion due to velocity field oscillations, E_{ij} stands for the dissipation of the stress into heat at the small eddies level, Π_{ij} refers to the pressure-strain interaction term and Ω_{ij} represents production of the turbulent stress by the system rotation. The one of the most important terms in the transport equation of the Reynolds stresses (5) is the pressure-strain interaction term; it describes interaction between velocity and pressure fluctuations. There are different variations of the RSM which depend on the approach that this term is described with.

3.1 Geometry

The rotor of the modelled wind turbine was based on the design experimentally tested by Kamoji [21]. The geometry of the turbine rotor was variant of the most generic shape, typical for the Savonius type wind turbine. Two semicylindrical shaped blades fastened with the two disks at the top and bottom side with overlapping gap between them. The schematic representation of the turbine rotor is presented in Fig. 1. The rotor diameter and the height denoted by *D* and *H* respectively, were equal to 208 mm. The overlap ratio of the rotor vanes (a) was set at 0.15 of the blade diameter. This overlap ratio was experimentally tested and suggested to be close to optimal for this type of Savonius turbine [21]. The thickness of the material forming the rotor blade was 2 mm, while the thickness of each disc at the top and bottom of the rotor was equal to the 10 mm. The diameters of top and bottom disks were equal to 1.1 times the diameter of the rotor.

The computational domain was defined as cuboid shaped body, conceptually mimicking the wind tunnel. The dimensions of the specific elements of the computational domain, as well as types of prescribed boundary conditions are presented in Fig. 2. The rotational movement of the turbine rotor was modelled using the Sliding Mesh

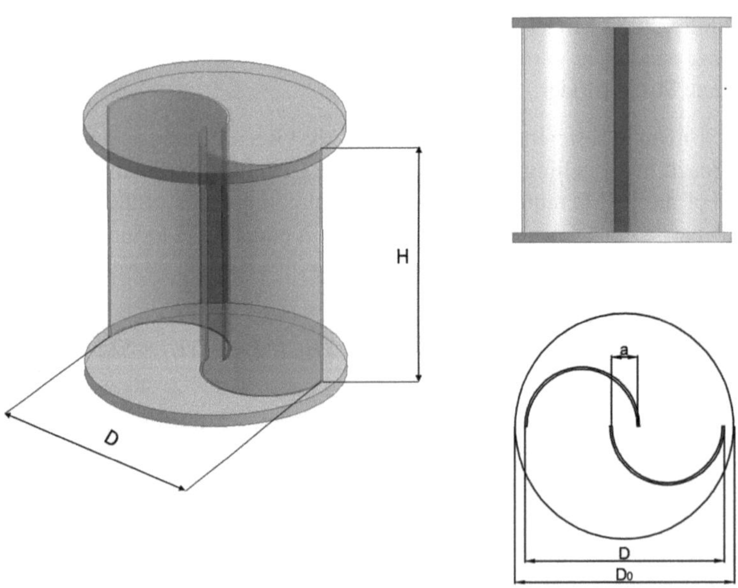

Fig. 1 Schematic view of the computational domain

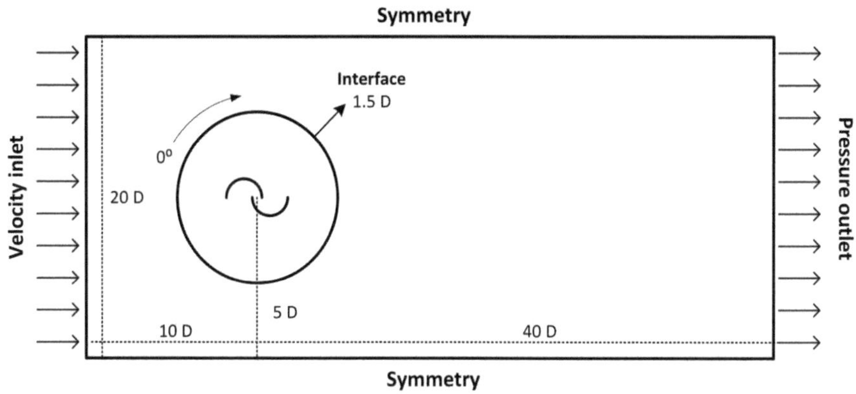

Fig. 2 Schematic view of the modelled Savonius wind turbine

method. Therefore, the computational domain was divided into the two separate sub-volumes: internal volume in cylindrical shape which can rotate around its axis and stationary surrounding. The domains were connected with the interface boundary conditions which ensures continuity between fields of the dependent variables. In order to limit the number of necessary mesh elements while retaining the high resolution of the numerical grid, the size of the domain was reduced to the upper half, with the symmetry boundary condition applied. The influence of such domain reduction was evaluated beforehand in terms of not only the turbine torque calculation but also the fluid flow prediction. As it was verified, the discrepancies between the full model and reduced model were negligible, there was no reason for doubling the necessary computational effort. The computational domain was spatially discretised with unstructured numerical mesh, primarily with the tetrahedral elements. In the wake region of the domain located behind the turbine rotor the numerical mesh the density of the mesh elements was purposely increased. For correct calculation of the physical phenomena in the close proximity of the turbine rotor, the boundary layer of prism type cells was applied. The size distribution of the numerical cells was prescribed in a manner for the y^+ parameter to not exceed the value of 2.5, with average value below 0.7. Finally, the total count of the mesh elements reached the number of 2.94×10^6. Considering relatively low values of the airflow velocity around the turbine rotor, fully segregated pressure-based solver was employed. All flow variables including the variables describing the turbulence and Reynolds Stresses were spatially discretised with the second order scheme, the pressure was discretised using staggered grid scheme (PRESTO! - pressure staggered grid method for arbitrary meshes). The pressure-velocity coupling in CFD solver was handled with the SIMPLE - Semi-Implicit Method for Pressure-Linked Equations algorithm. All elements of the numerical VAWT model, including the geometry developing, numerical mesh generation and proper calculation were conducted with the commercial Ansys software platform [22].

4 Results

The computations were carried out in transient regime; at the initial state it was assumed that the surrounding air is at rest and the wind starts blowing and turbine starts to rotate once the simulations is started. Therefore, in order to obtain reliable estimate of the turbine performance computations needs to be carried out until pseudo-steady state is reached. For the analysed wind turbine such a pseudo-steady state conditions were reached after 10 revolutions. The time step size for the temporal integration of the solution was chosen in such a way to limit the azimuthal increment of the turbine rotor within the one time step to 0.5 degree. At the inlet boundary condition constant wind velocity equal to 10 m/s was prescribed. Simulations of the operating Savonius wind turbine were conducted for five different operating conditions described by the Tip-Speed-Ratio parameter values equal to 0.1, 0.5, 0.8, 1.2 and 1.4 accordingly. With constant wind speed prescribed the rotational speed of the internal domain was calculated with the TSR formula:

$$TSR = \frac{\omega D}{2w_\infty} \tag{6}$$

The value of the turbine torque production was then averaged over two full rotations of the rotor - in order to minimise possible numerical instabilities. With the averaged value of the turbine torque, the averaged value of the turbine coefficient of power could be estimated:

$$C_p = \frac{2T_{avg}\omega}{\rho w_\infty^3 DH} \tag{7}$$

The performed computations involving the different turbulence models revealed several important issues (Fig. 3). Firstly, the $k - \epsilon$ realisable model poorly predicts performance of the presented VAWT turbine. The obtained results showed signifi-

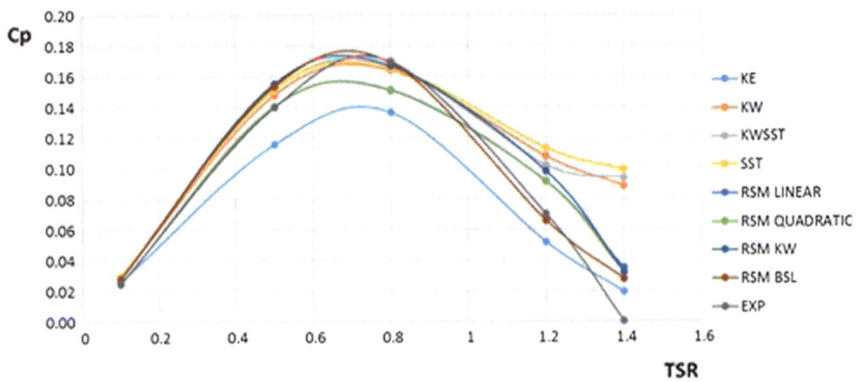

Fig. 3 Influence of turbulence modelling approach on the predicted wind turbine efficiency

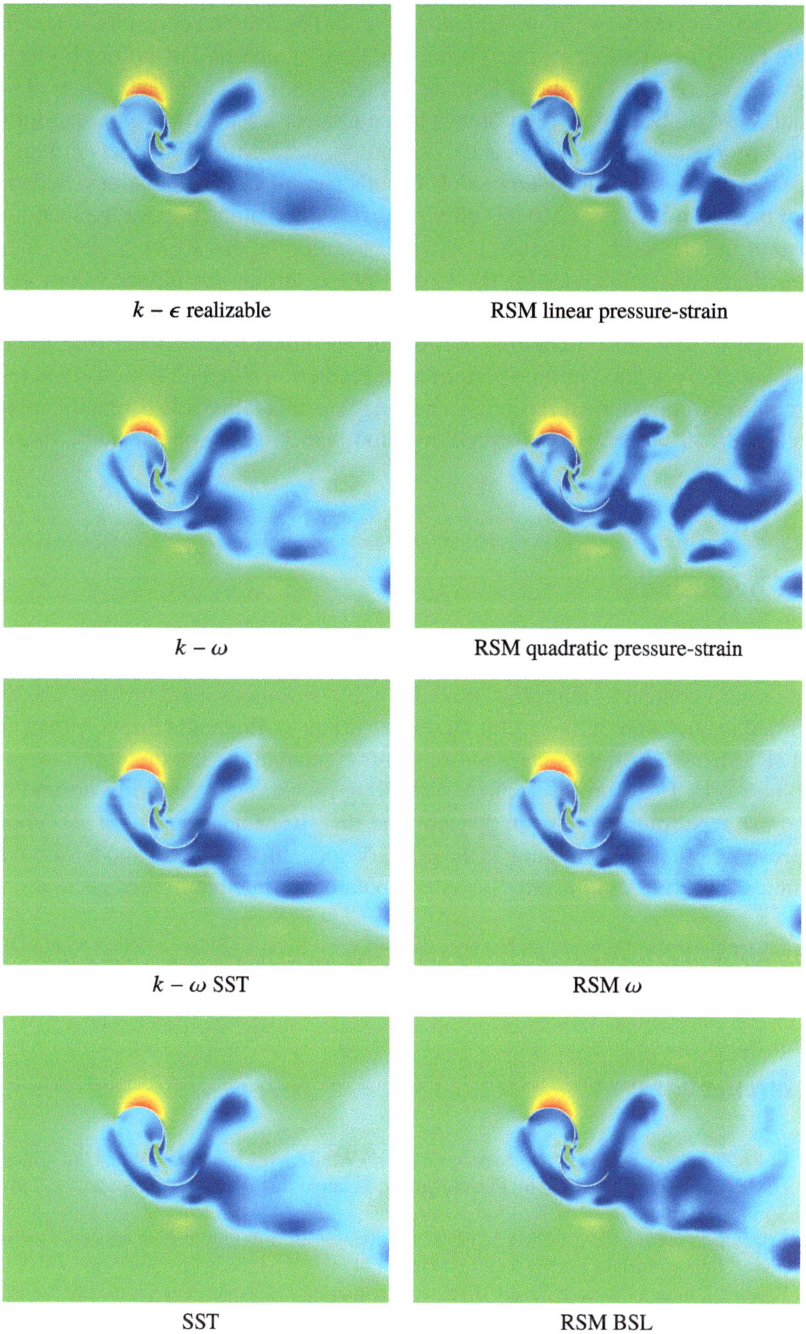

Fig. 4 Velocity field prediction with different turbulence models

cant discrepancies between model predictions and measurement results. The omega based family of turbulence models offer very good prediction capabilities for turbine operation range with Tip-Speed-Ratio below the value of 1. For higher values of rotational speed of the turbine rotor omega based models drastically overpredict turbine torque production. This is especially important, because these models are among most commonly recommended models for this type of CFD calculations. Secondly, all of the Reynolds Stress Models offer relatively good prediction for the mentioned range of turbine work. However, it is worth mentioning that RSM models can also correctly predict production of the torque by the turbine for higher values of the TSR parameter. Performed analysis suggests that the most suitable model for Savonius type wind turbine modelling is Baseline Reynolds Stress Model (Fig. 4). This model offers very good turbine performance prediction although it is costly in terms of required computational resources - calculations involving RSMs require approximately 30–40 percent more computing time than largely popular eddy viscosity models.

5 Conclusions

This work compares isotropic and non-isotropic eddy viscosity closure models performance in context of numerical modelling of a savonious wind turbine. The most accurate results were obtained for Baseline variant of Reynolds Stress Model. The BSL RSM model offers very good predicting capabilities for the entire range of turbine work. Therefore, if accurate modelling of Vertical Axis Wind turbine with Savonius type rotor is the main concern, the BSL RSM appears as the best choice. However, if efficient computing is the primary concern, $k - \omega$ SST model should be sufficient, offering significantly lower computational time but limited accuracy.

Acknowledgements The research has been supported by National Science Centre within OPUS scheme under contract UMO-2017/27/B/ST8/02298.

References

1. Burton T, Sharpe D, Jenkins N, Bossanyi E (2001) Wind energy handbook. Wiley Ltd, Chicester
2. Hau E (2006) Wind turbines. In: Fundamentals, technologies, application, economics, 2nd edn. Springer, Berlin
3. Ericsson S, Bernhoff H, Leijon M (2008) Evaluation of different turbine concepts for wind power. Renew Sustain Energy Rev 12:1419–1434
4. Pope K, Dincer I, Naterer GF (2010) Energy and exergy efficiency comparison of horizontal and vertical axis wind turbines. Renew Energy 35:2102–2113
5. Agren O, Berg M, Leijon M (2005) A time-dependent potential flow theory for the aerodynamics of vertical axis wind turbines. J Appl Phys 97:104913
6. Delgaire P, Engblom S, Ågren O, Bernhoff H (2009) Analytical solutions for a single blade in vertical axis turbine motion in two-dimensions. Eur J Mech B/Fluids 28:506–520

7. Islam MR, Mekhilef S, Saidur R (2013) Progress and recent trends of wind energy technology. Renew Sustain Energy Rev 21:456–468
8. Pope SB (2000) Turbulent flows. Cambridge University Press, Cambridge
9. Versteeg HK, Malalasekera W (2007) An introduction to computational fluid dynamics. The finite volume method, 2nd edn. Pearson Education Limited, Harlow
10. Wilcox DC (1998) Turbulence modelling for CFD. DCW Industries Inc., La Canada, California
11. Rolland S, Newton W, Williams AJ, Croft TN, Gethin DT, Cross (2013) Simulations technique for the design of a vertical axis wind turbine device with experimental validation. Appl Energy 111:1195–1203
12. Nasef MH, El-Askary WA, AbdEL-hamid AA, Gad HE (2013) Evaluation of Savonius rotor performance: static and dynamic studies. J Wind Eng Ind Aerodyn 123:1–11
13. D′ Alessandro V, Montelpare S, Ricci R, Secchhiaroli A (2010) Unsteady aerodynamics of a Savonius wind rotor: a new computational approach for the simulation of energy performance. Energy 35:3349–3363
14. Castelli MR, Englaro A, Benini E (2011) The Darrieus wind turbine: proposal for a new performance prediction model based on CFD. Energy 36:4919–4934
15. McTavish S, Feszty D, Sankar T (2012) Steady and rotating computational fluid dynamics simulations of a novel vertical wind turbine for small-scale power generation. Renew Energy 41:171–179
16. Rossetti A, Pavesi G (2013) Comparison of different numerical approaches to the study of the H-Darrieus turbines start-up. Renew Energy 50:7–19
17. Castelli MR, Dal Monte A, Quaresimin M, Benini E (2013) Numerical evaluation of aerodynamic and inertial contributions to Darrieus wind turbine blade deformation. Renew Energy 51:101–112
18. Almohammadi KM, Ingham DB, Ma L, Pourkashan M (2013) Computational fluid dynamics (CFD) mesh independency techniques for a straight blade vertical axis wind turbine. Energy 58:483–493
19. Zhou T, Rempfer D (2013) Numerical study of detailed flow field and performance of Savonius wind turbines. Renew Energy 51:373–381
20. Kacprzak K, Liskiewicz G, Sobczak K (2013) Numerical investigation of convectional and modified Savonius wind turbines. Renew Energy 60:578–585
21. Kamoji MA, Kedare SB, Prabhu SV (2008) Experimental investigations on single stage, two stage and three stage conventional Savonius rotor. Int J Energy Res 32:877–895
22. Ansys Fluent documentation 18.2 Release

The Standard and Counter-Rotating VAWT Performances with LES

Horia Dumitrescu, Alexandru Dumitrache, Ion Malael and Radu Bogateanu

Abstract Traditionally, the wind turbine performance is defined in terms of power extraction performance (expressed non-dimensionally as power coefficient, C_P, with its maximum value $C_{P_B} = 16/27$) while the turbine ability to start is normally ignored. Nevertheless, if a turbine cannot accelerate through start-up, its power extraction performance is severely limited, especially at low wind speeds. The criterion of starting behavior at relatively low Reynolds numbers, appropriate for the urban application, therefore offers another expectation to improve the overall performance concerning the period that the turbine needs to start might be achieved which might lead to a significant increase in energy turn-out. The work will focus upon vertical-axis machines of Darrieus type using an H-rotor in which the blades are straight and parallel to their axis of rotation. For the small such turbines, i.e. in low Reynolds number flows, some researchers have stated that the Darrieus-type turbine is inherently not self-starting. The concept of a vertical-axis counter-rotating rotor is used to overcome the starting drawback of small Vertical Axis Wind Turbines (hereafter VAWT). For this purpose, we attempt to simulate the flow around a Counter-Rotating VAWT (CR-VAWT) with Large Eddy Simulation (LES) and both starting behavior and power performance is outlined by comparing with an equivalent conventional turbine.

H. Dumitrescu · A. Dumitrache (✉)
"Gh. Mihoc-C. Iacob" Institute of Mathematical Statistics and Applied Mathematics,
Calea 13 Septembrie 13, 050711 Bucharest, Romania
e-mail: alex_dumitrache@yahoo.com

H. Dumitrescu
e-mail: dumitrescu.horia@yahoo.com

I. Malael
National Research and Development Institute for Gas Turbine, COMOTI,
Blvd. Iuliu Maniu 220, 061126 Bucharest, Romania
e-mail: ion.malael@comoti.ro

R. Bogateanu
INCAS – National Institute for Aerospace Research "Elie Carafoli",
Blvd. Iuliu Maniu 220, 061126 Bucharest, Romania
e-mail: bogateanu.radu@incas.ro

E. Ferrer and A. Montlaur (eds.), *Recent Advances in CFD for Wind and Tidal Offshore Turbines*, Springer Tracts in Mechanical Engineering,
https://doi.org/10.1007/978-3-030-11887-7_11

1 Introduction

Much work has been done to maximise the power extraction of wind turbines blades, but their starting characteristics have typically been ignored. However, wind turbines are also required to be self-starting and whilst blades designed for maximum power extraction can be optimised aerodynamically (thin airfoils with small drag coefficients), these blades often have poor starting performance. The experimental and computational studies of the starting process [1, 2] have pointed out a major problem with the aerodynamic optimization for both small horizontal and vertical axis wind turbines, in the sense that the optimized blades are the slowest ones to start. Furthermore, the good starting performance is generally more critical for small rather than large turbines, i.e., for small Reynolds numbers. Therefore, for the small turbines it is necessary to understand the starting process at small Reynolds numbers to allow a dual optimization by trading off power-production capability against starting performance, i.e. without negative performance band at low tip speed ratios. This can be easy to achieve for horizontal axis wind turbines because starting torque is generated mainly near the hub, whereas most power-producing comes from the tip region. Thus, simply increasing of the blade chord will increase its torque but will simultaneously increase its inertia.

However, both chord and twist distributions can be traded power coefficient off against starting time. Through optimizing the chord and twist in combination with the tip speed ratio an improvement of the power coefficient by 10% and a factor of 20 gain in starting time was reported [2].

The combination of two counter-rotating vertical axis wind turbines has been studied by many researchers with the task to improve the starting regime. A two counter rotating rotors configuration, Fig. 1b, is numerically investigated to determine the efficiency, the starting performance and to understand the physics of flow around the fluid-blade contact at low Reynolds numbers.

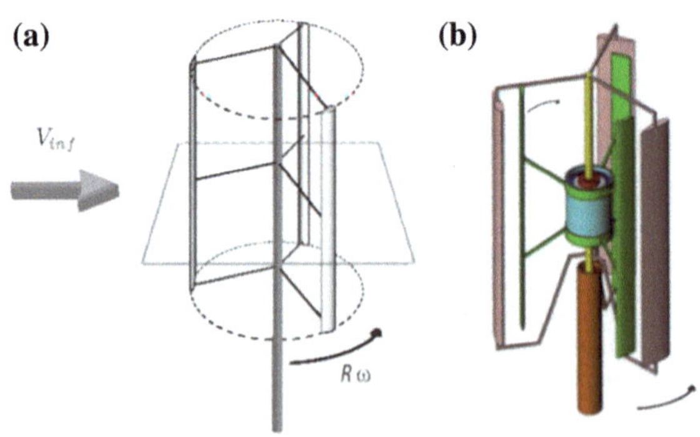

Fig. 1 **a** Standard straight blade VAWT; **b** Co-axis counter rotating-VAWT

2 The Starting of Small VAWTs

The solution of self-starting for small vertical rotors is more subtle and is connected to the existence of an elastic twist impact between incoming flow and blades produced at the critical Reynolds number, $\mathrm{Re}_{cr} \cong 10^5$ for a wind velocity of $V_{W\,\mathrm{min}} = 2.5$ m/s and a chord of 1 m. The elastic starting impact in critical Reynolds number flows involves a twist potential energy of impacted fluid, of molecular thermal nature, which can be exploited for improving both self-starting at low tip speed ratios ($TSR \sim 1.0$) and power production at higher TSR [3, 4].

The current operation of VAWTs involves a succession of collisions between incoming flow and a blade number, at any running regime. The impact energy is an intrinsic twist energy of fluid in contact with blade which was previous ignored in energy budget of turbines, important especially for small turbines.

The collision/impact is a process of momentum and/or kinetic energy transfer from incoming flow to blades depending on the impact velocity: the momentum of small velocities is used for starting, while the kinetic energy of higher velocities is used for improving power production. This property of elastic collisions known as the "Ω"-shape momentum-energy invariance is used for optimization of small VAWTs.

The twist/torsional potential so-called internal energy of fluid is found in the form of twisted vorticity of fluid in wall-bounded flows where its state depends on the well-known Reynolds number. The threshold of free twist energy is the critical Reynolds number, $\mathrm{Re}_{c,cr} \cong 10^5$ at a minimum wind velocity of $V_{W\,\mathrm{min}} = 2.5$ m/s and a chord of 1 m.

Figure 2 shows the spectral distributions of energy for the starting impact at $\mathrm{Re}_{\mathrm{cr}}$, which indicates a share of 20% from half of impact energy, $\rho V_W^2/2$, as the internal twist energy available for improving both starting and power production performance. The starting contribution was ignored in previous analyses.

The free internal/twist energy is transported in the fluid stream under impact in the lee of blades by a standing wave system. Lee waves are waves which travel through the fluid with a velocity equal and opposite to that of the flow so that they remain stationary relatively to the obstacle/blade producing them. The flow-blade impact is associated with the presence of some ordered ensembles of vortices of various types, with complex dynamics (see Fig. 5). Practically, such vortex coherent structures near blades promote a starting with short idling time, so-called "whirlwind" starting, which can be achieved by a "properly" sizing. Aerodynamically, these vortical concentrated structures induce drag reduction at low TSR, also called drag dynamic stall, and lift increasing (or lift dynamic stall) at higher TSR [5].

The contra-rotating turbine is a solution for obtaining such boundary concentrated vorticities which will be described in the sequel.

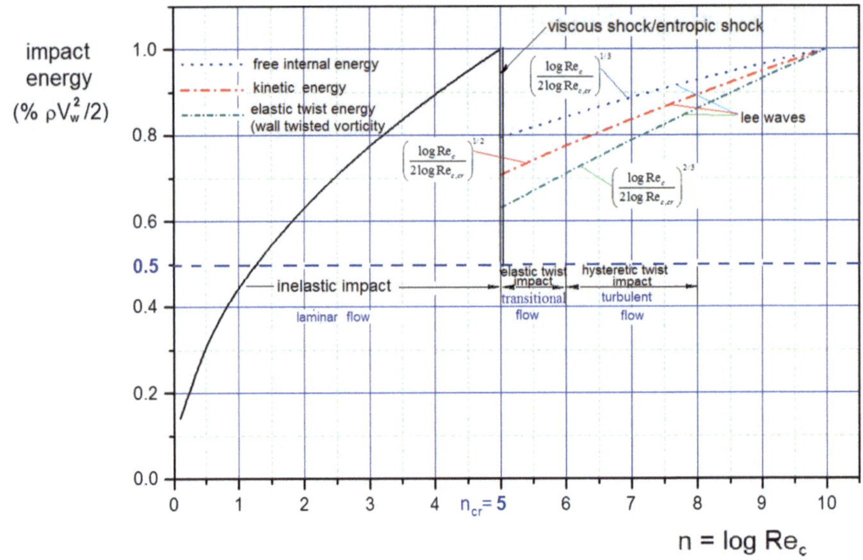

Fig. 2 Spectral distribution of impact available energy $\left(n_{cr} = \log Re_{c,cr} = \log 10^5\right)$

3 Numerical Analysis for a Co-axis Counter-Rotating VAWT

The combination of two VAWTs has been studied by many researchers. The usual combination is between fast/lift Darrieus and slow drag Savonius types [6, 7]. The configuration, Darrieus-Savonius was used to improve Darrieus wind turbine at starting. It is well known that the Savonius rotor creates high torque able to self-start at low wind speeds, but is relatively low in efficiency rating.

The Darrieus rotor has not a self-starting capabilities, but has much higher efficiency than the Savonius rotor. The combination of rotors increases the total power of the turbine in lower wind speed removing the starting drawback of conventional Darrieus turbines.

In this section the concept of counter-rotating rotor as a solution capable for self-start without compromising its power performance at higher TSR, is analysed. The co-axis counter-rotating VAWT (CR-VAWT) have two H-type rotors with three straight blades both using the NACA 0021 airfoil.

In the Table 1 the geometric parameters of this wind turbine are shown. The parameters of a conventional turbine for comparison are similar to the out rotor.

To solve the large separated unsteady-flow, numerical simulations are conducted using the incompressible Navier-Stokes equations and the Menter sub-grid scale model for turbulence.

Table 1 The geometric parameters of the CR-VAWT

Parameters	Value		Units
	Turbine_in	Turbine_out	
Airfoil	NACA 0021	NACA 0021	–
Blade chord	0.3	0.5	m
Turbine diameter	1.8	2.4	m
Turbine height	3.6	3.6	m
Number of blades	3	3	–
Design velocity	10		m/s
Design power	2000		W

3.1 Numerical Methods

The numerical investigations of the flow around of the CR-VAWT utilizes commercial CFD codes based on Unsteady Reynolds Averaged Navier-Stokes (URANS) and Large Eddy Simulation (LES) standard methods which can describe both the unsteady flow and turbulence effects in a transitional flow $Re_c = 4 \times 10^5$. The URANS method associated with the Shear Stress Turbulence (SST) model [8] and a LES filtering method [9] can capture the vorticity concentrations involved in transitional wall-bounded flows. To understand the complex unsteady flow around a CR-VAWT the URANS including SST turbulence model have been used due to its accurate predictions of wind turbine performance proved successfully in previous work [10].

3.2 Numerical Setup

For the numerical simulation a 2D domain has been defined and this has been split in four subdomains, two stationary and two rotating.

The rotational subdomain contains the blades of the inner and outer turbine, while the stationary subdomains are represented by the environment around the wind turbine and inner stationary domain, Fig. 3a.

The grid is generated by means of ICEM CFD software using a blocking function, to have a structured grid with hexa-elements, Fig. 3b. To resolve the boundary-layer flow, the y^+ value has been set lower than 1, and the growth ratio of elements was set to 1.05.

The CFD methods consist in solving the incompressible Navier-Stokes equations by means of the commercial software ANSYS Fluent where the required input parameters of this CFD case are given in Table 2. The CFD simulations of CR-VAWT rely on a mesh model which works by selecting a rotational speed set for the turbine and performing several tests for determining the nominal point of working turbine.

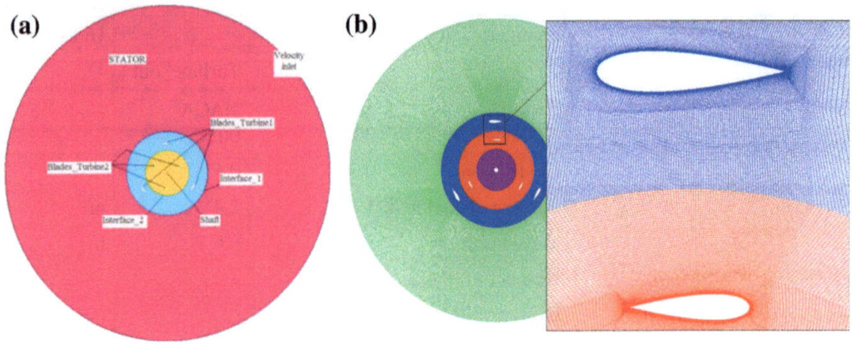

Fig. 3 Computational setup for three straight-bladed CR-VAWT: **a** computational domain; **b** 2D mesh structure

Table 2 Ansys fluent case setup

Models	Solver	Pressure based	Unsteady	2D
	Viscous model	k-ω SST		
Materials	Air	Density, constant		
Operating conditions	Pressure 101,325 (Pa)			
Boundary condition	Inlet	Velocity inlet $V_x = 13$ m/s		
	Blades	Wall		
	Shaft	Wall		
	Interfaces	Interface rotor-stator and rotor-rotor		
	Rotors	Mesh motion		
	Stators	Stationary		
Solve	Controls	Solution	Courant nr = 5	Discretization 2nd order upwind
	Initialize	Inlet, Velocity = 13 m/s		
	Monitors	Residuals	10^{-3}	
		Force	Momentum coefficient	
	Iterate	1800 steps	0.001 s time step size	
Report	Reference values	Inlet	Length = turbine radius	

3.3 Numerical Results

Figure 4 shows both C_T—torque and C_P—power coefficients for the CR-VAWT by comparison with the performance an equivalent conventional Darrieus turbine. The C_T and C_P curves show some expected similarities with the Gaussian bell-shape distribution of loading produced by successive flow-blade collisions during the running of turbine. Thus, in contrast to the Darrieus turbine, the CR-VAWT preserves a near periodic behaviour of the torque coefficient both at low TSR and high TSR showing a continuous-lift driven mode for the whole running period, Fig. 4a. Figure 4b shows that the free intrinsic energy from successive flow-blade collisions is able to completely remove the negative performance band ($C_P < 0$) at low TSRs and even to increase efficiency at higher TSRs.

The dynamic stall phenomenon is an inherent regime of the operation of a VAWT at low tip speed ratios ($TSR < 3$) where for $TSR \sim 1$ the starting behavior is crucial for the rotor enters its steady operating state, Fig. 4b. At low speed ratios, in the

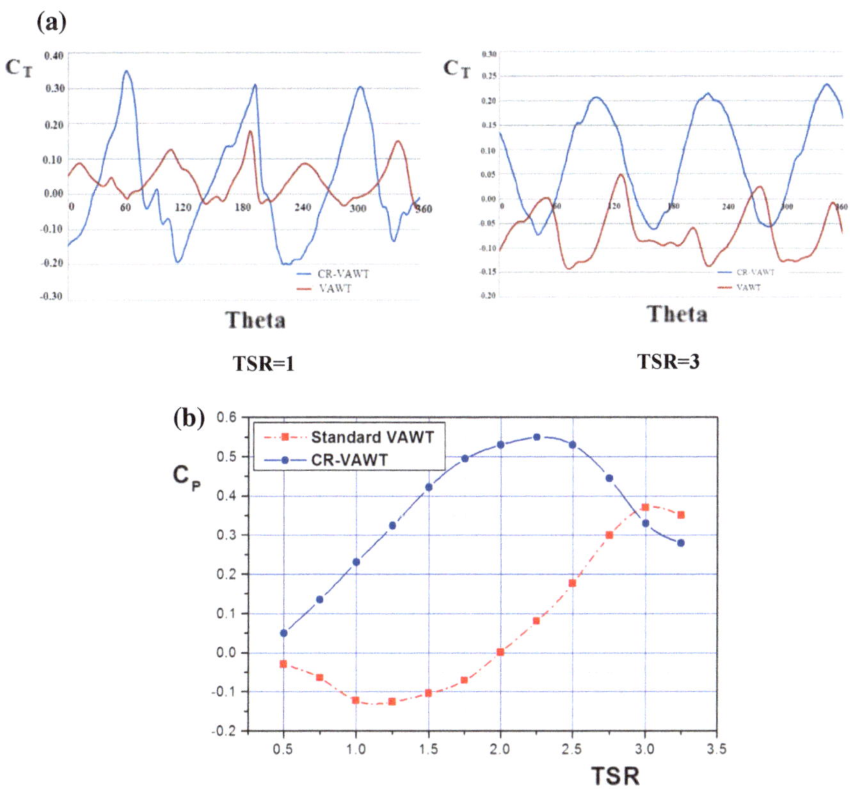

Fig. 4 Performance of a two co-axis CR-VAWT against the VAWT: **a** Torque coefficient—C_T; **b** Power coefficient—C_P

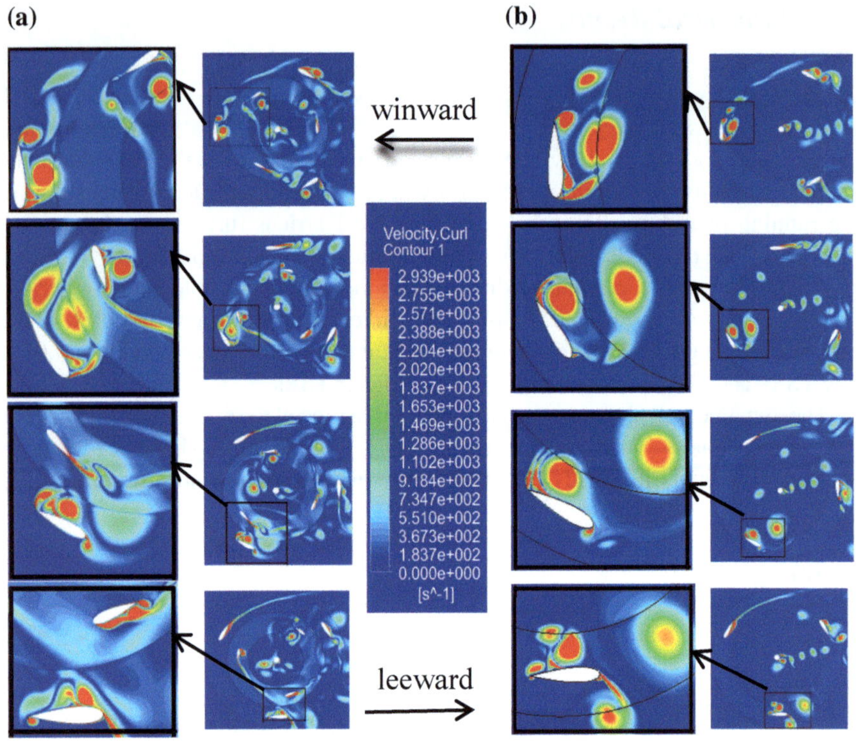

Fig. 5 Vorticity magnitude contours for TSR = 1: **a** CR-VAWT; **b** VAWT

range of chord Reynolds number $\mathrm{Re}_c = 10^5 - 10^6$, the near wake is captured inside the rotor and its shed vorticity is accumulated and concentrated at the windward quarter ($0° < \theta \leq 90°$) of the rotation. Then in the leeward quarter ($90° < \theta \leq 180°$) of the rotation, the concentrated vortical structures detach from the airfoil (centrifugal effect) and induce the velocity/pressure field around airfoil resulting in modified aerodynamic forces, i.e. the significant drag reduction. This is the drag dynamic stall event promoting the continuous thrust-producing and self-starting [5]. Figure 5a illustrates for CR-VAWT the comparatively evolution of the leading-edge counterclockwise concentrated vorticity which remains attached to the upper side of airfoils in the whole leeward quarter of the rotation producing the drag dynamical stall and fully removing of negative performance band at low TSRs. In contrast to the CR-VAWT, the conventional VAWT in leeward direction exhibits large zones of separated flow on the upper side of airfoil, where the concentrated vorticity is suddenly dissipated and the drag force increases, Fig. 5b. The concentrated vorticity in the vicinity of airfoil is a kind of inertia (fluctuating wall pressure) which must trade off against the aerodynamic performances of blades (profile type, chord, aspect ratio, blade number) for dual purpose of maximizing both power coefficient and starting performance, i.e. the shortest "idling time".

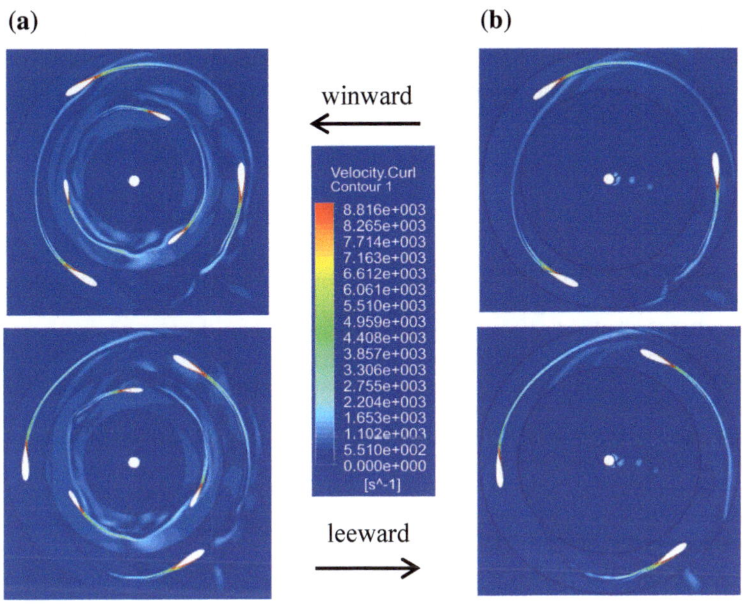

Fig. 6 Vorticity distributions around a CR-VAWT and VAWT, predicted by LES for TSR = 3: **a** CR-VAWT; **b** VAWT

At high tip speed ratio ($TSR \sim 3$) the two velocity fields are uncoupled and the outer rotor is running without compromising the wind turbine performance, Fig. 6. Additionally, it can be seen that the twisted flow between rotors has a beneficial effect on the performance of the main/outer rotor, a power gain of more 10% is found.

4 Conclusions

To understand the unsteady large separated flow around a counter-rotating VAWT, numerical simulations were conducted with LES. The numerical results showed that the concept of counter-rotating turbine offers self-start capabilities at low TSR concomitantly with improving of power production at high TSR. On the other hand, the effect of drag reduction, so called drag dynamic stall, induced by successive flow-blade collisions, wholly removed the negative power coefficient band at low TSR. As a result of lift dynamic stall at TSR = 2 the power coefficient was rapidly increasing with more 10% than the efficiency of an equivalent conventional turbine. From an economic standpoint and social "going green" phenomena the concept of self-starting counter-rotating VAWT proves to be a viable solution for small wind turbine market and weak wind environments.

An important outcome of theoretical nature shows that the counter-rotating concept obeys the "Ω" momentum-energy invariance which governs the elastic collision phenomenon producing a "whirlwind" starting and improved efficiency.

References

1. Wright AK, Wood DH (2004) The starting and low speed behaviour of a small horizontal axis wind turbine. J Wind Eng Ind Aerodyn 92:1265–1279
2. Wood DH (2004) Dual purpose design of small wind turbines for optimal starting and power extraction. Wind Eng 23(1):15–21
3. Hill N, Dominy R, Ingram G, Dominy J (2009) Darrieus turbines: the physics of self-starting. Proc IMechE Part A J Power Energy 223:21–28
4. Dumitrescu H, Dumitrache A, Frunzulica F, Pal A, Turbatu V (2013) TORNADO concept and realisation of a rotor for small VAWTs. INCAS Bulletin 5(3):69–75
5. Dumitrescu H, Cardos V, Malael I (2015) The physics of starting process for vertical axis wind turbines. In: Ferrer E, Montlaur A (eds) CFD for wind and tidal offshore turbines, Chap 7. Springer Tracts in Mechanical Engineering, pp 69–81
6. Menet J-L (2004) A double-step Savonius rotor for local production of electricity: a design study. Renew Energy 29:1843–1862
7. Tjiua W, Marnotob T, Mata S, Ruslana MH, Sopiana K (2015) Darrieus vertical axis wind turbine for power generation II: challenges in HAWT and the opportunity of multi-megawatt Darrieus VAWT development. Renew Energy 75:560–571
8. Menter FR, Esch T, Kubacki S (2002) Transition modelling based on local variables. In: 5th international symposium on turbulence modeling and measurements, Spain
9. Franke J, Hellsten A, Sclunzen H, Carissimo B (2007) Best practice for the CFD simulation of flows in the urban environment: COST Action 732 quality assurance and improvement of microscale meteorological models: Meteorological Inst.
10. Mălăel I, Drăgan V, Vizitiu G (2015) The vertical axis wind turbine efficiency evaluation by using the CFD methods. Appl Mech Mater 772:90–95

A High-Order Finite Volume Method for the Simulation of Phase Transition Flows Using the Navier–Stokes–Korteweg Equations

Abel Martínez, Luis Ramírez, Xesús Nogueira, Fermín Navarrina and Sofiane Khelladi

Abstract In this work, we employ the Navier–Stokes–Korteweg system of equations for the simulation of phase transition flows. This system belongs to the diffuse interface models, in which both phases are separated by a non-zero thickness interface where the properties vary continuously. The key idea of these methods is the ability to use the same set of equations for the entire computational domain, regardless of the phase of the fluid. However, these methods lead to a system of equations with high-order derivatives, which are difficult to discretize and solve numerically. Here, we propose the use of a high-order Finite Volume method, FV-MLS, for the resolution of the Navier–Stokes–Korteweg equations. The method uses Moving Least Squares approximations for the direct and accurate discretization of higher-order derivatives, which is particularly suitable for simulations on unstructured meshes. In this work, we show two numerical examples in which the interface is set to interact with great changes in the properties, in order to demonstrate the robustness of the method.

1 Introduction

This work is focused on water cavitation, which consists on the vaporization of a liquid due to big drops of the pressure. The interaction between turbomachinery and a liquid produces an interchange of forces, distributed between momentum and pressure. A decline the pressure could lead to a phase change on the fluid, producing

A. Martínez (✉) · L. Ramírez · X. Nogueira · F. Navarrina
Universidade da Coruña, Group of Numerical Methods in Engineering,
Campus de Elviña, 15071 A Coruña, Spain
e-mail: abel.mdiaz@udc.es

L. Ramírez
e-mail: luis.ramirez@udc.es

S. Khelladi
Laboratoire de Dynamique des Fluides, Arts et Métiers ParisTech,
151 Boulevard de l'Hôpital, 75013 Paris, France

E. Ferrer and A. Montlaur (eds.), *Recent Advances in CFD for Wind and Tidal Offshore Turbines*, Springer Tracts in Mechanical Engineering,
https://doi.org/10.1007/978-3-030-11887-7_12

that, in a small space, liquid and gas phases of the same substance may coexist and interact with each other. Cavitation is a known problem in turbomachinery, where production of small bubbles and their later collapse lead to peaks of pressure that may damage the structure of the blades, and thus affect its correct functioning.

To simulate a fluid that can be in either gas or liquid states with phase change, we use the Navier–Stokes–Korteweg (NSK) equations. This system belongs to the diffuse interface models [1], which are characterized for separating different phases with a non-zero width layer where the properties of the fluid vary continuously. The diffuse interface models apply the same system of equations to all the domain, avoiding algorithms to track the different phases of the fluid and imposing different conditions to each one. The main drawback of these models is the high-order derivatives that appear on the formulation [2].

2 Navier–Stokes–Korteweg Multiphase Flow Model

There are some numerical solutions of the NSK equations in literature, employing the Finite Difference method [3], the Discontinuous Galerking method [4] and Isogeometric analysis [5]. However, very few solutions are found on unstructured meshes, due to the difficulty to compute accurately the high-order derivatives in the Korteweg tensor. In this work, we propose the use of a high-order Finite Volume method based on Moving Least Squares approximations (FV-MLS) [6–11] to discretize the NSK equations.

2.1 Isothermal Navier–Stokes–Korteweg Equations

The system of equations consists in the compressible Navier–Stokes equations with the Korteweg tensor, which is added to model the capillarity forces on the interface [2]. This term introduces derivatives of order three. Following [5], the isothermal version of the Navier–Stokes Korteweg equations can be written in non-dimensional form as system of conservation laws as

$$\frac{\partial \boldsymbol{u}}{\partial t} + \nabla \cdot \left(\boldsymbol{F}^H - \boldsymbol{F}^E - \boldsymbol{F}^K \right) = \boldsymbol{S} \tag{1}$$

where $\boldsymbol{u}$ is the vector of variables, $\boldsymbol{F}^H$ is the inviscid flux, $\boldsymbol{F}^E$ is the viscous flux, $\boldsymbol{F}^K$ is the Korteweg flux and $\boldsymbol{S}$ is a source term. In this work, the source term is used for the simulation of gravitational forces in the example in Sect. 4.2.

$$u = \begin{Bmatrix} \widehat{\rho} \\ \widehat{\rho v_x} \\ \widehat{\rho v_y} \end{Bmatrix}. \tag{2}$$

$$\boldsymbol{F}_x^H = \begin{pmatrix} \widehat{\rho}\widehat{v_x} \\ \widehat{\rho}\widehat{v_x}^2 + \widehat{P} \\ \widehat{\rho}\widehat{v_x}\widehat{v_y} \end{pmatrix} \quad \boldsymbol{F}_y^H = \begin{pmatrix} \widehat{\rho}\widehat{v_y} \\ \widehat{\rho}\widehat{v_x}\widehat{v_y} \\ \widehat{\rho}\widehat{v_y}^2 + \widehat{P} \end{pmatrix} \tag{3}$$

$$\boldsymbol{F}_x^E = \frac{1}{Re}\begin{pmatrix} 0 \\ 2\partial_x\widehat{v_x} - \frac{2}{3}\left(\partial_x\widehat{v_x} + \partial_y\widehat{v_y}\right) \\ \left(\partial_y\widehat{v_x} + \partial_x\widehat{v_y}\right) \end{pmatrix} \quad \boldsymbol{F}_y^E = \frac{1}{Re}\begin{pmatrix} 0 \\ \left(\partial_y\widehat{v_x} + \partial_x\widehat{v_y}\right) \\ 2\partial_y\widehat{v_y} - \frac{2}{3}\left(\partial_x\widehat{v_x} + \partial_y\widehat{v_y}\right) \end{pmatrix} \tag{4}$$

$$\boldsymbol{F}_x^K = \frac{1}{We}\begin{pmatrix} 0 \\ \left(\widehat{\rho}\Delta\widehat{\rho} + \frac{1}{2}\left|\nabla\widehat{\rho}\right|^2 - \partial_x\widehat{\rho}\partial_x\widehat{\rho}\right) \\ -\left(\partial_x\widehat{\rho}\partial_y\widehat{\rho}\right) \end{pmatrix} \quad \boldsymbol{F}_y^K = \frac{1}{We}\begin{pmatrix} 0 \\ -\left(\partial_y\widehat{\rho}\partial_x\widehat{\rho}\right) \\ \left(\rho\Delta\widehat{\rho} + \frac{1}{2}\left|\nabla\widehat{\rho}\right|^2 - \partial_y\widehat{\rho}\partial_y\widehat{\rho}\right) \end{pmatrix} \tag{5}$$

$$\boldsymbol{S} = -\rho^*\boldsymbol{f}^* = -\rho^*\begin{Bmatrix} 0 \\ 0 \\ f_y \end{Bmatrix}. \tag{6}$$

where $\widehat{\rho}$ is the dimensionless density, $\widehat{P}$ is the dimensionless pressure and $\widehat{v} = (\widehat{v_x}, \widehat{v_y})^T$ the dimensionless velocity field.

The thickness of the interfaces is related to the Weber number We, which defines the ratio between the surface tension and the inertia of the fluid. Following [5], we scale the Weber number with an arbitrary length L_0 and a characteristic length of the mesh h as

$$h = max\sqrt{A_i} \qquad We = \frac{L_0^2}{h^2} \qquad L_0 = 1 \tag{7}$$

where A_i is the area of an element i in the mesh. We relate the Reynolds number to the Weber number through the following expression

$$Re = \alpha\sqrt{We} \qquad \alpha = 2 \tag{8}$$

Moreover, the Bond number Bo indicates the relative importance of gravitational forces compared to the surface tension. The Bond number can be defined as

$$Bo = We|\boldsymbol{f}^*| \tag{9}$$

2.2 Equation of State

For the system of equations to be completed, it is required to define an equation of state that represents the relation between pressure and density for both liquid and gas phases. This equation of state must be continuous through both phases, in accordance with the idea of diffuse interface models. The van der Waals equation of state [12, 13] is commonly used, although there are other possibilities. The van der Waals equation of state reads in non-dimensional form as

$$\widehat{P} = \frac{8\widehat{T}\widehat{\rho}}{27(1-\widehat{\rho})} - \widehat{\rho}^2 \tag{10}$$

with

$$\widehat{T} = \frac{T}{T_{crit}} = \frac{27R}{8ab}T \qquad \widehat{P} = \frac{1}{ab^2}P \qquad \widehat{\rho} = \frac{1}{b}\rho \tag{11}$$

where the parameters a and b are constants depending on the substance and R is the ideal gas constant.

For a temperature below the critical temperature, the van der Waals equation of state produces a local maximum and a minimum under the saturation curve, as it is plotted in Fig. 1. This figure can be interpreted as a vapor phase on the left side of the spinodal curve and a liquid phase on the right side, joined together continuously through the interface.

The saturation curve plotted in Fig. 1 represents the region in which the thermodynamic equilibrium between both phases is produced, therefore, they can coexist stably [14].

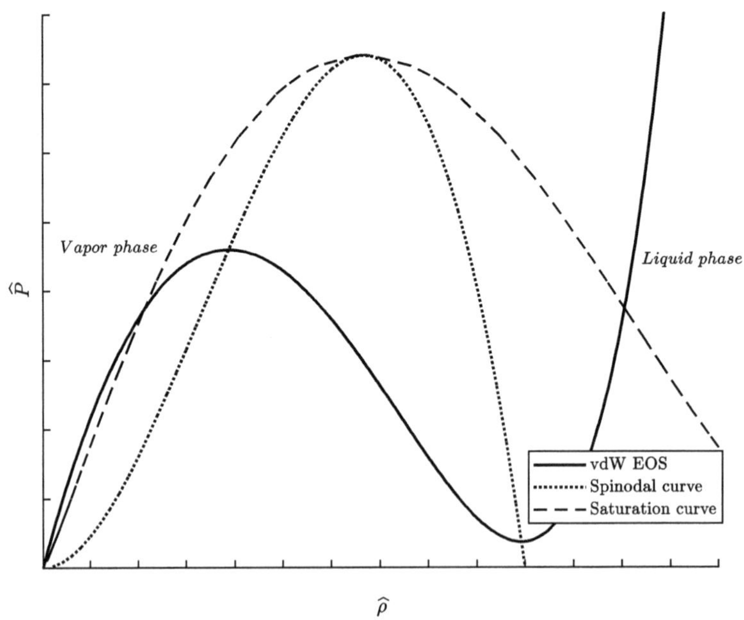

Fig. 1 Representation of the van der Waals equation of state

3 The FV-MLS Method

In order to discretize the system of equations described in (1), we propose the use of the FV-MLS method [6–11]. We use the MLS method to obtain the derivatives of order three found in the Korteweg term (Sect. 2.1) as it is an especially useful technique to use in unstructured meshes.

Given a function $\boldsymbol{u}(\boldsymbol{x})$ with its value $\boldsymbol{u}_I$ at the cell centroid $\boldsymbol{x}_I$, so that $\boldsymbol{u}_I = \boldsymbol{u}(\boldsymbol{x}_I)$, Moving Least Squares approximates the value of the variable at a point $\boldsymbol{x}$ with a function $\boldsymbol{u}^h(\boldsymbol{x})$ as

$$u(\boldsymbol{x}) \approx \boldsymbol{u}^h(\boldsymbol{x}) = \sum_{i=1}^{m} p_i(\boldsymbol{x})\alpha_i(\boldsymbol{z})\mid_{z=\boldsymbol{x}} = \boldsymbol{p}^T(\boldsymbol{x})\boldsymbol{\alpha}(\boldsymbol{z})\mid_{z=\boldsymbol{x}} \tag{12}$$

where $\boldsymbol{p}^T$ is a m dimension basis of functions, which for one dimensional case it can be written as

$$\boldsymbol{p}(x) = (1, x, x^2, ..., x^{m-1}) \tag{13}$$

$\boldsymbol{\alpha}(\boldsymbol{z})\mid_{z=x}$ is a vector of coefficients that must be determined. The values of the coefficients are obtained minimizing a function $J(\boldsymbol{\alpha}(\boldsymbol{z})\mid_{z=\boldsymbol{x}})$, which is the weighted mean square error committed when $\boldsymbol{u}(\boldsymbol{x})$ is approximated as $\boldsymbol{u}^h(\boldsymbol{x})$. The function $J(\boldsymbol{\alpha}(\boldsymbol{z})\mid_{z=\boldsymbol{x}})$ can be defined as

$$J(\boldsymbol{\alpha}(\boldsymbol{z})\mid_{z=\boldsymbol{x}}) = \int_{y\in\Omega_x} W(\boldsymbol{z}-\boldsymbol{y},\lambda)\mid_{z=\boldsymbol{x}} \left[u(\boldsymbol{y}) - \boldsymbol{p}^T(\boldsymbol{x})\boldsymbol{\alpha}(\boldsymbol{z})\mid_{z=\boldsymbol{x}}\right]^2 d\Omega_x \tag{14}$$

where $W(\boldsymbol{z}-\boldsymbol{y},\lambda)\mid_{z=\boldsymbol{x}}$ is a weighted function known as Kernel [7]. This function weights the values of the different points that take place in the interpolation according to the distance to the center $\boldsymbol{z} = \boldsymbol{x}$. The parameter λ is the smoothing length that defines the size of the support.

Minimizing Eq. (14) and obtaining an expression of the coefficients $\boldsymbol{\alpha}(\boldsymbol{z})\mid_{z=\boldsymbol{x}}$, we can replace it in Eq. (12) and obtain an expression of the form

$$\boldsymbol{u}^h(\boldsymbol{x}) = \sum_{j=1}^{n_I} N_j(\boldsymbol{x})\boldsymbol{u}_j \tag{15}$$

where $N_I(\boldsymbol{x})$ are a set of *shape functions* associated to the centroids and n_I are the neighboring nodes that participates in the approximation of the point $\boldsymbol{x}$. The set of nodes for each point $\boldsymbol{x}$ is referred as the *stencil*.

Integrating the system of conservation laws in a control volume I and applying the divergence theorem to the flux terms we get the integral form

$$\int_{\Omega_I} \frac{\partial \boldsymbol{u}}{\partial t} d\Omega + \int_{\Gamma_I} \left(\boldsymbol{F^H} - \boldsymbol{F^E} - \boldsymbol{F^K}\right) \cdot \boldsymbol{n}\, d\Gamma = \boldsymbol{0} \tag{16}$$

where Ω_I and Γ_I are respectively the area and the perimeter of each control volume I and the term $\boldsymbol{n} = (n_x, n_y)^T$ is the exterior normal vector.

Introducing the approximated variable $\boldsymbol{u}^h$ in (16), we obtain

$$\int_{\Omega_I} \frac{\partial \boldsymbol{u}^h}{\partial t} d\Omega + \int_{\Gamma_I} \left(\boldsymbol{F}^{hH} - \boldsymbol{F}^{hE} - \boldsymbol{F}^{hK} \right) \cdot \boldsymbol{n} \, d\Gamma = \boldsymbol{0} \tag{17}$$

The FV-MLS method splits in the formulation the hyperbolic and elliptic fluxes and treats them differently, this way the method acknowledges their different nature. Reordering the hyperbolic and the elliptic type of fluxes, we obtain

$$\int_{\Omega_I} \frac{\partial \boldsymbol{u}^h}{\partial t} d\Omega + \int_{\Gamma_I} \Theta(\boldsymbol{u}^{hb+}, \boldsymbol{u}^{hb-}) \, d\Gamma - \int_{\Gamma_I} \left(\boldsymbol{F}^{hH} + \boldsymbol{F}^{hE} - \boldsymbol{F}^{hK} \right) \cdot \boldsymbol{n} \, d\Gamma = \boldsymbol{0} \tag{18}$$

For the computation of the hyperbolic flux, we employ a *"broken"* reconstruction, which generates a continuous reconstruction of the variable $\boldsymbol{u}(\boldsymbol{x})$ inside each cell although discontinuous at the cell interfaces. This reconstruction is obtained using high-order Taylor series expansions from the cell centroid. At the integration points at the cell interfaces, the numerical flux is computed using the Rusanov Riemann solver [15] with the Li and Gu's fix for all-speed flows [16], that can be expressed as

$$\boldsymbol{\Theta}_{i+\frac{1}{2}} = \frac{1}{2}(\boldsymbol{F}^{hH+} + \boldsymbol{F}^{hH-}) \cdot \boldsymbol{n} - \frac{1}{2} S^{+} \boldsymbol{f(M_l)} \Delta(\boldsymbol{u}) \tag{19}$$

with

$$S^{+} = max(|\boldsymbol{v}^{+}| + c^{+}, |\boldsymbol{v}^{-}| + c^{-}) \tag{20}$$

$$\boldsymbol{f(M_l)} = \boldsymbol{I} \left\{ \begin{array}{l} 0 \\ min(\frac{|v|}{c}, 1) \\ min(\frac{|v|}{c}, 1) \end{array} \right\}. \tag{21}$$

where c is the speed of sound, $|\boldsymbol{v}|$ is the modulus of the velocity vector at the integration point and $\Delta(\boldsymbol{u}) = (\boldsymbol{u}^{hb+} - \boldsymbol{u}^{hb-})$.

For the elliptic fluxes, MLS is centered at the integration points, getting a highly accurate and continuous reconstruction.

4 Numerical Results

In this section, we show two different applications of the method explained.

4.1 Coalescence of Two Vapor Bubbles

In this example, we establish two vapor bubbles immersed on a liquid with different radii. Due to the pressure that the larger bubble exerts, the smaller bubble will shrink during the time of the simulation. When the bubbles are placed close enough, they will eventually merge to form a single and larger bubble, which will evolve to the typically expected circular shape. For this example, the computational domain is a square $\Omega = [0, 1] \times [0, 1]$, with periodic boundary conditions in all directions. We place their centers at $O_1 = (0.40, 0.50)^T$ and $O_2 = (0.75, 0.50)^T$. Their radii are set to be $R_1 = 0.25$ and $R_2 = 0.10$ respectively. The initial condition for the density is established employing an hyperbolic tangent profile to model the interfaces (Eq. (22)) and the velocity field is set to be null, as it is shown in [5].

$$\widehat{\rho}(x, t = 0) = 0.10 + 0.25\left[tanh\left(\frac{d_1(x) - R_1}{2}\sqrt{We}\right) + tanh\left(\frac{d_2(x) - R_2}{2}\sqrt{We}\right)\right]$$

$$\widehat{v}_x, \widehat{v}_y = 0 \tag{22}$$

where $d_i(x)$ refers to the Euclidean distance between $\boldsymbol{x}$ and O_i, with $i = 1, 2$. We use a mesh of 256×256 elements. The temperature is set at $\widehat{T} = 0.85$. In Fig. 2, we show the time evolution of the density field, where it is shown the initial condition, the merge of the vapor bubbles and the final circular bubble. The interface that separates the two different phases is clearly identified in the figure. Moreover, in

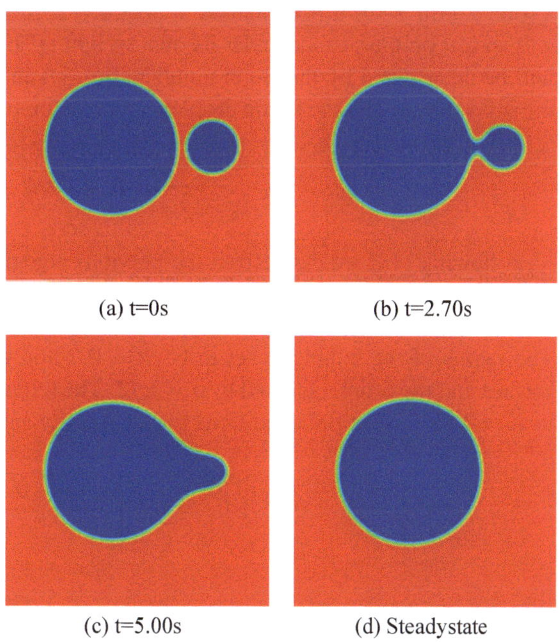

(a) t=0s (b) t=2.70s (c) t=5.00s (d) Steadystate

Fig. 2 Time evolution of the density field in the coalescence between two vapor bubbles

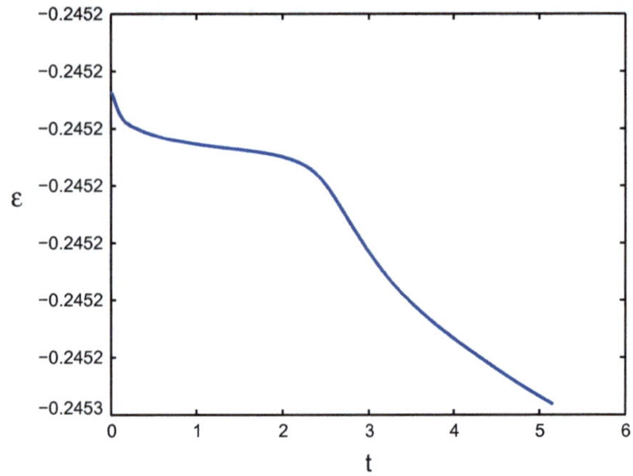

Fig. 3 Time evolution of the free energy in the coalescence of two vapor bubbles

Fig. 3 it is plotted the time evolution of the free energy, which is a monotonically decreasing function, as it is expected in this problem. Note that the variation of the slope corresponds with the merge of the vapor bubbles.

4.2 Liquid Droplet on a Flat Surface

The aim of this example is to simulate a liquid droplet in contact with a flat surface, modelled as a solid wall. For this simulation, we recreate the gravity force through a source term $\boldsymbol{f}^*$ as it is explained in Sect. 2.1. It is expected that the liquid droplet will spread on the surface under its own weight [17]. The final shape of the droplet will be determined by the Bond number (Eq. (9)) and the contact angle φ between the solid surface and the liquid droplet, which is imposed as a boundary condition as

$$-\frac{\nabla \rho}{\|\nabla \rho\|} \cdot \boldsymbol{n} = cos\varphi \tag{23}$$

Following [18] we discretized the previous equation using MLS approximation at the boundary. Again, our computational domain is a square $\Omega = [0, 1] \times [0, 1]$, with a wall in the bottom boundary, where no slip boundary condition is imposed. The radius of the bubble is set to be $R = 0.2$ and the contact angle between the surface and the droplet is fixed to $\varphi = \pi/2$. The temperature of the simulation is set to $\widehat{T} = 0.85$. The initial conditions for the density and velocity field are

$$\widehat{\rho}(x, t = 0) = 0.35 + 0.25tanh\left(\frac{\sqrt{(x - 0.5)^2 + y^2} - R}{2}\sqrt{We}\right)$$

$$\widehat{v}_x, \widehat{v}_y = 0 \tag{24}$$

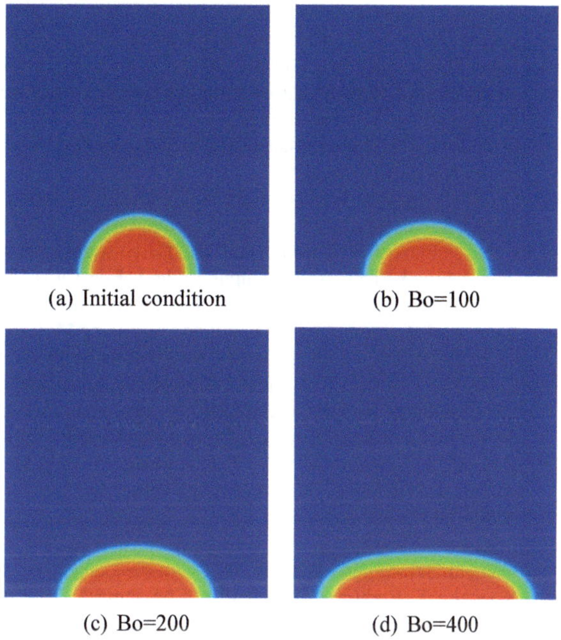

Fig. 4 Density field for a liquid droplet on a surface. The steady state simulations with different values of the Bond number are shown, all sharing the same initial condition shown in **a**

In Fig. 4, the initial condition of the problem studied and the steady state using different values of the Bond number are shown. An increase of the Bond number will increase the gravity effects with respect to the interface forces that hold the droplet, and thus the droplet will spread on the surface more noticeably.

Conclusions

In this work, we present the numerical resolution of the Navier–Stokes–Korteweg equations for the simulation of phase transition flow using Moving Least Squares approximations on a Finite Volume scheme. Moving Least Squares approximations are used to obtain a high-order reconstruction of the variables needed to compute the high-order derivatives that appear in the formulation. In the numerical examples, we show the ability to simulate the interaction between two interfaces of two different vapor bubbles and the interaction between the interface and wall-type boundary conditions.

References

1. Anderson DM, McFadden GB, Wheeler AA (1998) Diffuse-interface methods in fluid mechanics. Annu Rev Fluid Mech 30:139–165
2. Korteweg DJ (1901) Sur la forme que prennent les équations du mouvements des fluides si l'on tient compte des forces capillaires causées par des variations de densité consiérables mais

continues et sur la théorie de la capillarité dans l'hypothèse d'une variation continue de la densité. Archives Néerlandaises des Sciences Exactes et Naturelles. Series II 6:1–24
3. Jamet D, Torres D, Brackbill JU (2002) On the theory and computation of surface tension: the elimination of parasitic currents through energy conservation in the second-gradient method. J Comput Phys 182:262–276
4. Diehl D (2007) Higher order schemes for simulation of compressible liquid-vapor flows with phase change. PhD thesis
5. Gómez H, Hughes TJR, Nogueira X, Calo VC (2010) Isogeometric analysis of the isothermal Navier-Stokes-Korteweg equations. Comput Methods Appl Mech Eng 199:1828–1840
6. Cueto-Felgueroso L, Colominas I (2008) High-order finite volume methods and multiresolution reproducing kernels. Arch Comput Methods Eng 15(2):185–228
7. Cueto-Felgueroso L, Colominas I, Nogueira X, Navarrina F, Casteleiro M (2006) High order finite volume schemes on unstructured grids using Moving Least Squares construction, Application to shallow waters dynamics. Int J Numer Methods Eng 65:295–331
8. Cueto-Felgueroso L, Colominas I, Nogueira X, Navarrina F, Casteleiro M (2007) Finite volume solvers and moving least-squares approximations for the compressible Navier-Stokes equations on unstructured grids. Comput Methods Appl Mech Eng 196:4712–4736
9. Khelladi S, Nogueira X, Bakir F, Colominas I (2011) Toward a higher order unsteady finite volume solver based on reproducing kernel methods. Comput Methods Appl Mech Eng 200(29):2348–2362
10. Lancaster P, Salkauskas K (1981) Surfaces generated by moving least squares methods. Math. Comput. 37(155):141–158
11. Nogueira X, Ramírez L, Khelladi S, Chassaing J, Colominas I (2015) A high-order density-based finite volume method for the computation of all-speed flows. Comput Methods Appl Mech Eng 298:229–251
12. van der Waals JD (1873) On the continuity of the gaseous and liquid states. PhD thesis
13. van der Waals JD (1979) The thermodynamic theory of capillarity under the hypothesis of a continuous variation of density. J Stat Phys (Reprinted) 20:197–244
14. Serrin J (2008) The area rule for simple fluid phase transitions. J Elast 90:129–159
15. Rusanov VV (1962) The calculation of the interaction of non-stationary shock waves and obstacles. USSR Comput Math Math Phys 1(2):304–320
16. Li X-S, Gu C-W (2013) Mechanism of Roe-type schemes for all-speed flows and its application. Comput Fluids 86:56–70
17. Liu J, Gómez H, Evans JA, Hughes TJR, Landis CM (2013) Functional entropy variables: a new methodology for deriving thermodynamically consistent algorithms for complex fluids, with particular reference to the isotermal Navier-Stokes-Korteweg equations. J Comput Phys 248:47–86
18. Ramírez L, Nogueira X, Khelladi S, Chassaing J-C, Colominas I (2014) A new higher-order finite volume method based on moving least squares for the resolution of the incompressible Navier-Stokes equations on unstructured grids. Comput Methods Appl Mech Eng 278:883–901

An a Posteriori Very Efficient Hybrid Method for Compressible Flows

Javier Fernández-Fidalgo, Xesús Nogueira, Luis Ramírez and Ignasi Colominas

Abstract In this work we present a framework for a high-order hybrid method made up of an explicit finite-difference scheme and a member of the Weighted Essentially Non-Oscillatory (WENO) family. A new a posteriori switching criterion is developed based on the Multidimensional Optimal Order Detection (MOOD) method. The schemes tested here are chosen to illustrate the process, we select non-standard fourth order finite difference and fifth order compact and non-compact WENO, but any other combination of central finite differences and upwind schemes could be used as well. In this work we present a one-dimensional and a two-dimensional case to illustrate the speed, accuracy and shock-capturing properties of the proposed schemes.

1 Introduction

When performing an aerodynamic flow simulation of a wind turbine, since the Mach number under working conditions is typically under M = 0.3, reaching its peak value at the tips of the blades, the incompressible fluid hypothesis is used. However, if a more profound study wants to be developed, an all-speed compressible flow solver is required. Taking into account the compressibility effects of the fluid means that the density is no longer constant, large gradients in both density and/or pressure may occur for higher values of the Mach number, the incompressible fluid assumption is no longer valid, and thus the compressible Euler or Navier–Stokes equations should be used. When these equations are used, it is possible the existence of non-smooth regions in the flow, where the accuracy of the scheme is greatly reduced or may cause the simulation to blow up if they are not treated with special care. Several techniques have been explored in order to achieve high-order results while dealing

J. Fernández-Fidalgo (✉) · X. Nogueira · L. Ramírez · I. Colominas
Universidade da Coruña, Group of Numerical Methods in Engineering,
Campus de Elviña, 15071 A Coruña, Spain
e-mail: javier.fernandez1@udc.es

X. Nogueira
e-mail: xnogueira@udc.es

E. Ferrer and A. Montlaur (eds.), *Recent Advances in CFD for Wind and Tidal Offshore Turbines*, Springer Tracts in Mechanical Engineering,
https://doi.org/10.1007/978-3-030-11887-7_13

with these kind of phenomena with the consequential increase of computational cost. In this work we explore a new hybrid technique based on the a posteriori detection paradigm that combines a central Finite Difference (FD) scheme and a Weighted Essentially Non-Oscillatory (WENO) scheme. The present methodology is not only restricted to the schemes that are presented hereunder, but it could also be applied to any combination of central finite difference and WENO schemes regardless of their order of accuracy and/or compactness.

The key idea is to combine a scheme that can deal with the non-smooth parts of the flow (such as shocks, or contact discontinuities), with a scheme that is fast and accurate but does not have any shock-handling capabilities. The former scheme will only be used on those areas where the latter is unable to produce a quality solution. Thus, the accurate and reliable detection of the problematic zones plays an important role.

To that matter, several FD/WENO hybrid schemes have been proposed in the literature, such as the attempts made by Costa and Don [1] and Pirozzoli [2] to name a few. These approaches rely on an a priori type of detection, meaning that the shocks are predicted from one time-step to the next. As explained by Hu et al. [3], the main drawback of these approaches is that the shock locations are related to a priori guesses, and some efficiency can be lost due to over-detection of problematic zones.

In this work, we develop an a posteriori detection criterion based on the Multi-dimensional Optimal Order Detection (MOOD) method. The reader is referred to [4, 5] for further details on the MOOD paradigm. The main difference between the present approach and the one from Clain et al. [4, 5], is that we combine two similar order schemes with significant differences in terms of computational cost, while these authors use a single scheme ranging from arbitrary high-order to first order, so that the reconstruction order of the problematic cells is gradually downgraded up to first order in case all the other attempts have failed. This is the reason why we do not label the present approach as MOOD, since it is an a posteriori approach but the order of the chosen schemes is not downgraded significantly.

2 Two-Dimensional Compressible Euler Equations

We cast the Euler equations in generalized coordinates for a compressible fluid. Following [6], the general curvilinear transformation from the physical domain (x, y) to the computational space (ξ, η) is written in the following strong conservation form:

$$\frac{\partial \hat{\mathbf{U}}}{\partial t} + \frac{\partial \hat{\mathbf{F}}}{\partial \xi} + \frac{\partial \hat{\mathbf{G}}}{\partial \eta} = \mathbf{0} \tag{1}$$

where $\hat{\mathbf{U}}$ denotes the scaled vector of conservative variables, being the original vector $\mathbf{U} = (\rho, \rho u, \rho v, \rho E)^T$ and $\hat{\mathbf{F}}$ and $\hat{\mathbf{G}}$ are the generalized inviscid flux-vectors in the ξ and η directions, which can be expressed as

$$\hat{\mathbf{U}} = \frac{1}{J}\begin{pmatrix} \rho \\ \rho u \\ \rho v \\ \rho E \end{pmatrix} \quad \hat{\mathbf{F}} = \frac{1}{J}\begin{pmatrix} \rho \hat{U} \\ \rho u \hat{U} + \xi_x p \\ \rho v \hat{U} + \xi_y p \\ (\rho E + p)\,\hat{U} \end{pmatrix} \quad \hat{\mathbf{G}} = \frac{1}{J}\begin{pmatrix} \rho \hat{V} \\ \rho u \hat{V} + \eta_x p \\ \rho v \hat{V} + \eta_y p \\ (\rho E + p)\,\hat{V} \end{pmatrix} \tag{2}$$

where ρ is the density, u and v are the velocity components along the x and y axes, p is the pressure, and $E = \frac{p}{\rho(\gamma-1)} + \frac{1}{2}\left(u^2 + v^2\right)$ is the total energy per unit mass.

The ratio of specific heat coefficients is denoted as γ, which for an ideal, monatomic gas is $\gamma = 7/5$. The quantities ξ_x, ξ_y, η_x and η_y are known as the spatial metrics of the transformation. The inverse metrics (x_ξ, y_ξ, x_η and y_η) are obtained by analytic or numeric differentiation depending on whether the exact expression of the mapping between the physical space and the computational domain is available or not.

The Jacobian of the transformation is denoted as

$$J = \left|\frac{\partial (x, y)}{\partial (\xi, \eta)}\right| = \frac{1}{x_\xi y_\eta - x_\eta y_\xi} \tag{3}$$

and the following relations apply:

$$\begin{aligned} \xi_x &= J y_\eta \qquad & \eta_x &= -J y_\xi \\ \xi_y &= -J x_\eta \qquad & \eta_y &= J x_\xi \end{aligned} \tag{4}$$

$\hat{U}$ and $\hat{V}$ denote the contravariant velocity components, and are expressed as the inner product of the velocity vector $\mathbf{u} = (u, v)$ and the gradient of the metrics

$$\begin{aligned} \hat{U} &= \mathbf{u}^T \cdot \nabla\xi = \xi_x u + \xi_y v \\ \hat{V} &= \mathbf{u}^T \cdot \nabla\eta = \eta_x u + \eta_y v. \end{aligned} \tag{5}$$

To obtain the inverse metrics, a sixth-order compact finite difference scheme, as in [7], has been used. It is advised that the scheme used to calculate the metrics should be at least of the same order of accuracy as the schemes employed in the calculation. This rules out the option of calculating the inverse metrics via the (CR)WENO scheme because the use of biased stencils near the boundaries would ruin the global accuracy.

3 Numerical Schemes

The proposed hybrid scheme is based on the use of an explicit, Low Dispersion Finite Difference (LDFD) method for the smooth regions of the flow, and a Weighted Essentially Non-Oscillatory (WENO) scheme for the rest of the domain. In this

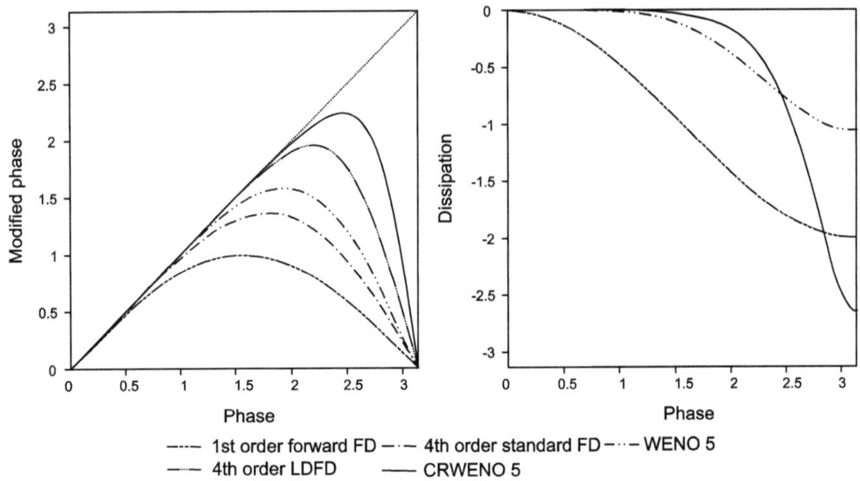

Fig. 1 Dispersion (left) and dissipation (right) diagrams

work we will analyze two different schemes obtained from the use of two different WENO schemes. Namely, a compact scheme (CRWENO 5) [8] and a non-compact scheme (WENO 5) [9]. It is worth noting that any other central difference and/or WENO scheme can be used in our approach regardless of its compactness or order of accuracy.

Dispersion and dissipation diagrams of all the schemes involved are shown in Fig. 1, where we also have included the curves for the first order forward and fourth order central finite differences for reference. Please note, that the LDFD scheme, and the fourth order standard finite differences, have no inherent dissipation.

3.1 Low Dispersion Finite Difference (LDFD) Scheme

In this work, the explicit, fourth order, centered, finite difference method proposed by Bogey and Bailly [10] has been chosen to be the base scheme. It has lower dispersion error compared to the standard fourth order finite differences at the expense of a wider stencil. The first derivative of the flux $\hat{\mathbf{F}}$ along the ξ-direction at some interior node, is computed as:

$$\left.\frac{\partial \hat{\mathbf{F}}}{\partial \xi}\right|_{i,j} \approx \sum_{p=0}^{N=5} s_p \left(\hat{\mathbf{F}}_{i+p,j} - \hat{\mathbf{F}}_{i-p,j}\right). \tag{6}$$

This scheme uses an eleven-point stencil, the coefficients s_p can be found in [10].

3.2 Filtering Schemes

Compact lowpass Padé filters proposed by Visbal and Gaitonde [11] are used to stabilize the LDFD scheme. The filtered value of a generic quantity χ, denoted as $\bar{\chi}$, is obtained by solving a tridiagonal system of the form:

$$\alpha_f \bar{\chi}_{i-1} + \bar{\chi}_i + \alpha_f \bar{\chi}_{i+1} = \sum_{n=0}^{N} \frac{a_n}{2} (\chi_{i+n} + \chi_{i-n}) . \tag{7}$$

The coefficients a_n depend on the order of the filter and can be found in [11] as well. A $2N$th order filter will have a stencil of $2N + 1$ points. The amount of dissipation introduced by the filter is controlled by α_f, which is retained as a free parameter. Varying the value of $\alpha_f \in (-0.5, 0.5)$ we can alter considerably the cutoff frequency of the low-pass filter and thus, the amount of dissipation introduced.

The filtering procedure is included to stabilize the numerical scheme for convection dominant problems. However, it is not enough to stabilize the LDFD numerical scheme in presence of shocks. Different strategies to stabilize a scheme in presence of shocks have been proposed in the literature. In [12] the interested reader can find a thorough overview on these strategies.

In this work we propose a different approach, related to the aforementioned hybrid approaches of Costa and Don [1] and Pirozzoli [2]. We use a WENO-family scheme to deal with non-smooth regions, as [13], but the region where the WENO scheme is used is determined *a posteriori* after the filtering step, only at regions where the solution is not acceptable.

3.3 WENO and CRWENO Schemes

For questions of brevity, and due to the popularity of the WENO family, the formulation will not be exposed here. We use a global Lax–Friedrichs flux splitting and the characteristic version, since it is explained in [9] that this formulation is more robust and produces less oscillatory results than the component-wise approach. The employed formulation of the schemes follows closely the one explained in [8], which is a mapped version proposed by [14] to the original schemes of Jiang and Shu [9].

3.4 Temporal Integration

In the present work, the total variation diminishing (TVD) three stage Runge Kutta (TVDRK3) [9] is used to advance the solution from time t^n to t^{n+1}:

$$\frac{\partial \mathbf{U}}{\partial t} = \mathbf{R} = -J\left(\frac{\partial \hat{\mathbf{F}}}{\partial \xi} + \frac{\partial \hat{\mathbf{G}}}{\partial \eta}\right) \tag{8}$$

$$\begin{aligned}
\mathbf{U}^{(1)} &= \mathbf{U}^n + \Delta t \mathbf{R}(\mathbf{U}^n) \\
\mathbf{U}^{(2)} &= \frac{3}{4}\mathbf{U}^n + \frac{1}{4}\mathbf{U}^{(1)} + \frac{1}{4}\Delta t \mathbf{R}(\mathbf{U}^{(1)}) \\
\mathbf{U}^{n+1} &= \frac{1}{3}\mathbf{U}^n + \frac{2}{3}\mathbf{U}^{(2)} + \frac{2}{3}\Delta t \mathbf{R}(\mathbf{U}^{(2)}).
\end{aligned} \tag{9}$$

4 A Posteriori HFDWENO and HFDCRWENO Schemes

4.1 Overview and Switching Criterion

Depending on the choice of the WENO scheme that will be combined with the LDFD scheme, we present two hybrid schemes called HFDWENO and HFDCRWENO respectively. In order to efficiently switch from the LDFD scheme to the (CR)WENO schemes, we have developed a criterion based on the a posteriori limiting paradigm [4, 5, 15, 16].

A sketch of the present approach is shown in Fig. 2. Based on the solution obtained on the previous Runge–Kutta step, at the beginning of each Runge–Kutta step a *candidate solution* $\mathbf{U}^\star$ is computed using the LDFD scheme. Visbal and Gaitonde's lowpass filtering is carried out only on the third substage (the third equation of (9)) of the Runge–Kutta scheme. A number of detectors are run on the candidate solution to check if it has some desirable properties (explained in the upcoming epigraph). If all the detectors are fulfilled, the solution at that point is acceptable, so no switch to the (CR)WENO scheme is needed, but if any of the detectors rejects the candidate solution, that point along with some stencil around it (see [17] for details), is recalculated with the WENO 5 or CRWENO 5 scheme followed by a positivity preservation technique.

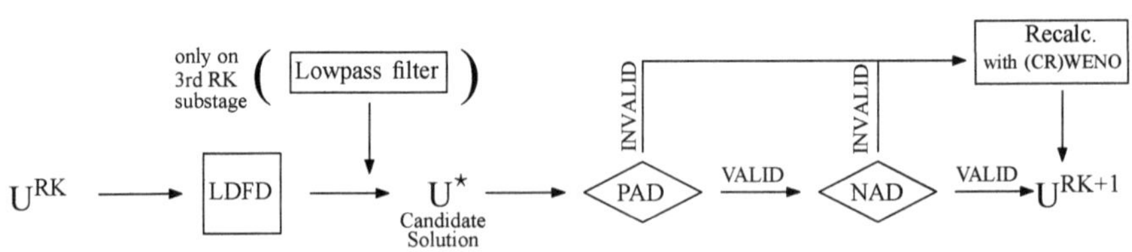

Fig. 2 Present a posteriori approach. $\mathbf{U}^{\mathrm{RK}}$ represents a Runge–Kutta substage and $\mathbf{U}^{\mathrm{RK}+1}$ represents the following substage of the Runge–Kutta algorithm

4.2 A Posteriori Detectors

An extensive description of the usually employed detectors in the MOOD paradigm can be consulted in [18], here we will only describe the ones employed in our formulation, sketched in Fig. 2.

Physical Admissible Detector (PAD): All points must have positive density and positive pressure at all times. In practice, this detector identifies points with negative or zero density and/or pressure values and also detects NaN values that usually arise from a failed calculation of a strong shock.
Numerical Admissible Detector (NAD) [15]: relaxed version of the Discrete Maximum Principle (DMP) [4]. It checks that no new extrema are created. It compares the candidate solution with the previously obtained solution in the previous RK substage.

$$\min_{\mathbf{y}\in\mathcal{V}_i}\left(\mathbf{U}^{\mathrm{RK}}(\mathbf{y})\right)-\delta \leqslant \mathbf{U}^{\star}(\mathbf{x}) \leqslant \max_{\mathbf{y}\in\mathcal{V}_i}\left(\mathbf{U}^{\mathrm{RK}}(\mathbf{y})\right)+\delta \tag{10}$$

$$\delta=\max\left(10^{-4}, 10^{-3}\cdot\left(\max_{\mathbf{y}\in\mathcal{V}_i}\left(\mathbf{U}^{\mathrm{RK}}(\mathbf{y})\right)-\min_{\mathbf{y}\in\mathcal{V}_i}\left(\mathbf{U}^{\mathrm{RK}}(\mathbf{y})\right)\right)\right) \tag{11}$$

where $\mathcal{V}_i$ represents the set of first neighbors of the point in consideration.

In this work, the NAD is checked on the full conservative variables vector as suggested by [5], although other authors [18] prefer to work just with the first component (density).

4.3 Recalculation of Flagged Points and Positivity Preservation

The points that do not fulfill any of the previously presented detectors, are recalculated along with a stencil around them. The stencil size coincides with that of the Finite Differences scheme employed to calculate the candidate solution (in this particular case 5 points to both sides of the flagged point in every direction), and this (CR)WENO region serves as an exclusion zone so that it is guaranteed that no oscillatory stencil containing the flagged point is used. Careful treatment of the borders of each (CR)WENO region must be taken, in order not to generate spurious oscillations. Detailed explanation on how all the above procedure is handled can be seen in [17].

Given that the (CR)WENO schemes are the last reconstruction procedure before entering the next Runge–Kutta substage, it is needed that they produce a solution with positive density and pressure. Based on the idea of [19], we completely replace the (CR)WENO computed flux by the first order Lax-Friedrichs flux (which is known to guarantee this property given that the CFL condition is not greater than 0.5) in case that pressure or density after the recalculation are not positive at some given point. The main difference with the method proposed in [19], is that they use a weighted average of the computed flux and the Lax-Friedrichs flux that introduces just the needed dissipation that guarantees the positivity, but our approach is faster in the computations and in the test cases we have addressed, the (CR)WENO schemes failed in very few cells and in rather scarce time steps, so this choice is justified. Please note that the positivity preservation is only applied to the recalculated regions, and is not involved in the process of generating the candidate solution.

5 Numerical Examples

To illustrate the capabilities of the hybrid methods to obtain a very similar solution to the original (CR)WENO schemes but using less computational time, two numerical examples are shown hereunder. Please note that the amount of saved time of the hybrid schemes (labeled hereafter as HFDWENO and HFDCRWENO respectively) compared to the original methods (labeled hereafter as WENO and CRWENO) is case- and scheme-dependent, obtaining greater time savings when the shock zones make up a small percentage of the whole domain. For the same test case, computationally costlier WENO-type schemes tend to produce better time savings than those less-demanding schemes. Please note that the higher the percentage of non-smooth parts over the total computational domain, the lower time savings will be obtained.

We present two cases, namely the 1D Sod shock tube and a 2D flow past a circular cylinder on a fully curvilinear mesh. For brevity, on Table 1 we present the computational times for the conventional and hybrid schemes, as well as the saved time expressed in percentage for the two mentioned cases.

5.1 1D Case. Sod Shock Tube

This is a well-known benchmark [20] which takes place in the one-dimensional interval [0, 1] with the initial condition given by:

$$(\rho, u, p) = \begin{cases} (1.000, 0.000, 1.000) & \text{if } x < 0.5 \\ (0.125, 0.000, 0.100) & \text{if } x \geq 0.5. \end{cases} \tag{12}$$

The CFL condition is set to 0.5. No time variation of the conservative variables has been specified at the left and right boundaries, which is appropriate for the time interval under interest. The simulation is run until $t = 0.17$. A 4th order filter with $\alpha_f = 0.45$ is employed.

This case requires no positivity preservation technique, since no points get negative density or pressure values throughout the simulation. We can see that the NAD

Table 1 Time comparative for the cases presented in Sects. 5.1 and 5.2

Case 5.1	CFL = 0.5	Total time (s)		Saved time (%)	Total time (s)		Saved time (%)
	Nodes	WENO	HFDWENO		CRWENO	HFDCRWENO	
	50	0.016	0.016	0.00	0.063	0.016	74.60
	100	0.016	0.015	6.25	0.250	0.063	74.80
	200	0.094	0.032	65.96	0.937	0.140	85.06
	400	0.359	0.110	69.36	3.609	0.312	91.35
	800	1.375	0.344	74.98	14.453	0.812	94.38
	1600	5.375	1.313	75.57	59.375	2.375	96.00
	3200	22.844	5.375	76.47	243.344	8.516	96.50
Case 5.2	CFL = 0.5	Total time (s)		Saved time (%)	Total time (s)		Saved time (%)
	Nodes	WENO	HFDWENO		CRWENO	HFDCRWENO	
	50 × 100	395.68	293.672	25.78	2989.64	1313.28	56.07
	100 × 200	2811.43	1651.21	41.27	20903.53	5796.92	72.27
	200 × 400	22885.87	12668.03	44.65	175499.047	34683.71	80.24

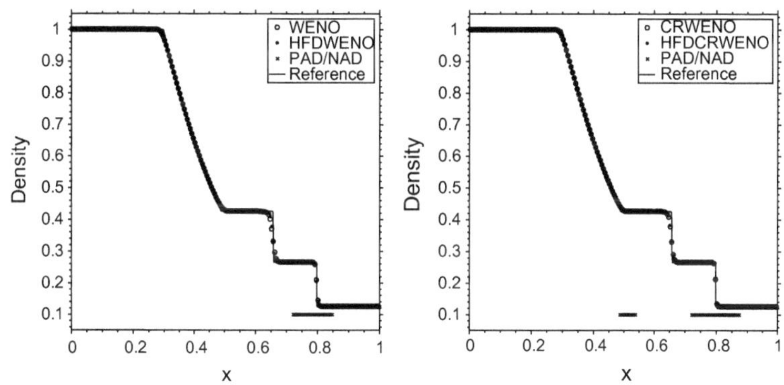

Fig. 3 A comparison of the hybrid schemes with respect to their classical counterparts is shown for a 200 node mesh for the final time of the simulation $t = 0.17$. The zones where the NAD is activated at the last time step are also shown. The reference solution has been computed with the WENO scheme on a 3200 node mesh

is activated around the shock location at $x = 0.8$ on both hybrids, and on the FDCR-WENO scheme there is also a small occurrence around $x = 0.5$ that disappears when the mesh is gradually refined. In Table 1 we show the computational times required by the schemes and we quantify the amount of saved time for this particular case. In Fig. 3, the obtained results are shown.

Given the special configuration of this case, a great reduction in computational time is achieved while obtaining a very similar solution to the original schemes.

5.2 2D Case. Mach 3 Supersonic Flow Past a Cylinder

We present a fully curvilinear case where the supersonic flow past a circular cylinder of unit radius is simulated. The mapping between the computational domain and the physical one, is expressed as [9]:

$$\begin{aligned} x &= -\left(R_X - (R_X - 1)\,\xi\right)\cos\left(\theta\,(2\eta - 1)\right) \\ y &= \quad \left(R_Y - (R_Y - 1)\,\xi\right)\sin\left(\theta\,(2\eta - 1)\right) \end{aligned} \tag{13}$$

where $R_X = 3$, $R_Y = 6$ and $\theta = 5\pi/12$. The computational domain ranges from 0 to 1 in both variables ξ and η. Solid wall boundary condition is applied on $\xi = 0$, inflow is applied at $\xi = 1$, and for $\eta = 0$ and $\eta = 1$ zeroth order extrapolation is used. Although this is a steady-state problem, we run the simulation to a total time of $t = 15$ to be able to compare the results of the schemes. This time is enough for the schemes to converge to a stationary solution. A 4th order filter with $\alpha_f = 0.45$ is employed. The positivity preserving algorithm is seldom activated in the early stages of the simulation for the CRWENO and its hybrid version.

In Table 2 the ratio of the stagnation pressure p_0 over the free-stream pressure p_∞ is shown, along with the stand-off distance non-dimensionalized by the diameter of the cylinder. In Fig. 4 we can see that all the schemes yield similar results. In Table 1 we show the results for both the hybrid schemes and the original ones and we can see that the time savings are again quite remarkable.

Table 2 Mach 3 supersonic flow past a cylinder. Resulting parameters for the 200 × 400 mesh

Scheme	p_0/p_∞	Stand-off distance/D	Notation
WENO	12.046	0.347	
HFDWENO	12.013	0.347	
CRWENO	12.056	0.344	
HFDCRWENO	12.051	0.344	
Reference value [21]	12.061	—	

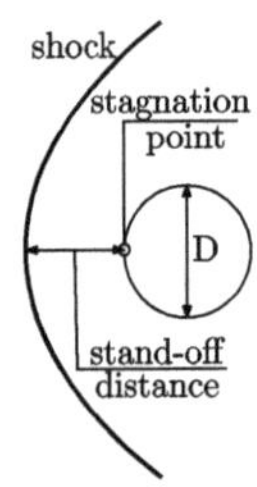

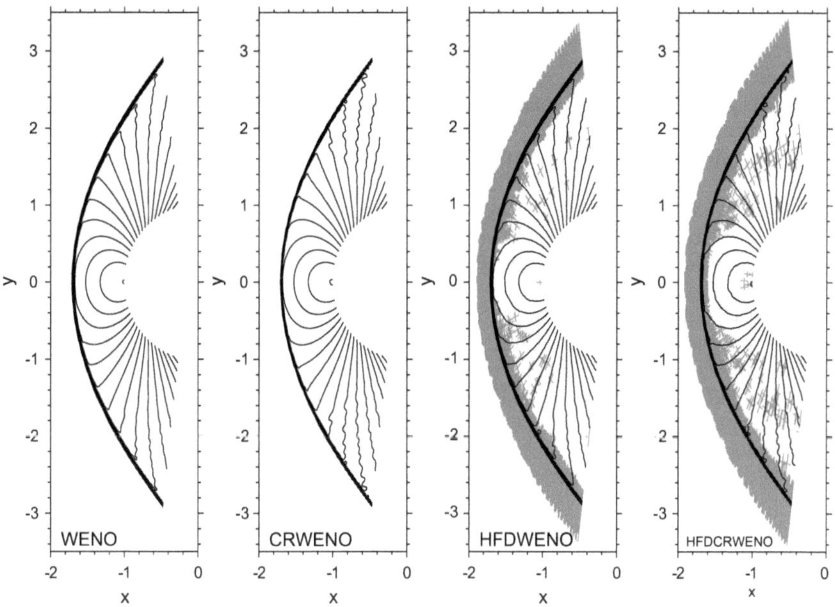

Fig. 4 Obtained results for the 200 × 400 mesh. The points where the PAD/NAD are activated at the last time step are also shown in gray for the hybrid schemes. 20 equally spaced pressure contours between 0.60 and 8.60 at time $t = 15$. Only the relevant part of the domain is shown

6 Conclusions

In this work we have presented a new high-order hybrid finite-difference scheme. It combines the speed of the central finite differences with the shock-handling capabilities of the WENO family schemes using an a posteriori switching criterion. The methodology presented here, can be extended to any combination of central and upwind schemes. It yields very accurate results while improving substantially the computational times of the original schemes.

References

1. Costa B, Don WS (2007) High order hybrid central-WENO finite difference scheme for conservation laws. J Comput Appl Math 204(2):209–218
2. Pirozzoli S (2002) Conservative hybrid compact-WENO schemes for shock-turbulence interaction. J Comput Phys 178(1):81–117
3. Hu XY, Wang B, Adams NA (2015) An efficient low-dissipation hybrid weighted essentially non-oscillatory scheme. J Comput Phys 301:415–424
4. Clain S, Diot S, Loubère R (2011) A high-order finite volume method for systems of conservation laws—multi-dimensional optimal order detection (MOOD). J Comput Phys 230(10):4028–4050

5. Diot S, Clain S, Loubère R (2012) Improved detection criteria for the multi-dimensional optimal order detection (MOOD) on unstructured meshes with very high-order polynomials. Comput Fluids 64:43–63
6. Vinokur M (1974) Conservation equations of gasdynamics in curvilinear coordinate systems. J Comput Phys 14(2):105–125
7. Visbal MR, Gaitonde DV (1999) High-order-accurate methods for complex unsteady subsonic flows. AIAA J 37(10):1231–1239
8. Ghosh D, Baeder JD (2012) Compact reconstruction schemes with weighted eno limiting for hyperbolic conservation laws. SIAM J Sci Comput 34(3):A1678–A1706
9. Jiang GS, Shu CW (1996) Efficient implementation of weighted ENO schemes. J Comput Phys 126(1):202–228
10. Bogey C, Bailly C (2002) A family of low dispersive and low dissipative explicit schemes for computing aerodynamic noise. In: 8th AIAA/CEAS aeroacoustics conference exhibit. American Institute of Aeronautics and Astronautics
11. Visbal MR, Gaitonde DV (1998) High-order schemes for Navier-Stokes equations: algorithm and implementation into FDL3DI. Air Force Research Lab Wright-Patterson, Air Vehicles Directorate
12. Johnsen E, Larsson J, Bhagatwala AV, Cabot WH, Moin P, Olson BJ, Rawat PS, Shankar SK, Sjögreen B, Yee HC, Zhong X, Lele SK (2010) Assessment of high-resolution methods for numerical simulations of compressible turbulence with shock waves. J Comput Phys 229(4):1213–1237
13. Adams NA, Shariff K (1996) A high-resolution hybrid compact-ENO scheme for shock-turbulence interaction problems. J Comput Phys 127(1):27–51
14. Henrick AK, Aslam TD, Powers JM (2005) Mapped weighted essentially non-oscillatory schemes: achieving optimal order near critical points. J Comput Phys 207(2):542–567
15. Dumbser M, Zanotti O, Loubère R, Diot S (2014) A posteriori subcell limiting of the discontinuous Galerkin finite element method for hyperbolic conservation laws. J Comput Phys 278:47–75
16. Nogueira X, Ramírez L, Clain S, Loubère R, Cueto-Felgueroso L, Colominas I (2016) High-accurate SPH method with multidimensional optimal order detection limiting. Comput Methods Appl Mech Eng 310:134–155
17. Fernández-Fidalgo J, Nogueira X, Ramírez L, Colominas I (2018) An a posteriori, efficient, high-spectral resolution hybrid finite-difference method for compressible flows. Comput Methods Appl Mech Eng 335:91–127
18. Figueiredo J, Clain S (2015) Second-order finite volume mood method for the shallow water with dry/wet interface. In: SYMCOMP
19. Hu XY, Adams NA, Shu CW (2012) Positivity-preserving flux limiters for high-order conservative schemes. J Comput Phys 242:169–180
20. Sod GA (1978) A survey of several finite difference methods for systems of nonlinear hyperbolic conservation laws. J Comput Phys 27(1):1–31
21. Liepmann HW, Roshko A (1959) Elements of gasdynamics. Wiley, New York. ISBN: 978-0486419633

MIX
Papier aus verantwortungsvollen Quellen
Paper from responsible sources
FSC® C105338
FSC
www.fsc.org

If you have any concerns about our products,
you can contact us on
ProductSafety@springernature.com

In case Publisher is established outside the EU,
the EU authorized representative is:
Springer Nature Customer Service Center GmbH
Europaplatz 3, 69115 Heidelberg, Germany

Printed by Libri Plureos GmbH
in Hamburg, Germany